Breakthrough Improvement with QI Macros and Excel®

About the Author

Jay Arthur teaches people how to eliminate the three silent killers of productivity and profitability—delay, defects, and deviation—using the Magnificent Seven Tools of Lean Six Sigma.

He excels at using the 4 percent of knowledge and tools that deliver over 50 percent of results. He has helped companies from health care to manufacturing save millions of dollars. He has found that using QI Macros and Microsoft Excel can help solve seemingly impossible problems quickly and easily.

Mr. Arthur is the author of *Lean Six Sigma Demystified* (McGraw-Hill) and six other books on Lean and Six Sigma. He is also the creator of QI Macros for Excel, an add-in that leverages the power of Excel to draw all of the charts and graphs needed for breakthrough improvement.

Breakthrough Improvement with QI Macros and Excel®

Finding the *Invisible* Low-Hanging Fruit

Jay Arthur

New York Chicago San Francisco
Athens London Madrid
Mexico City Milan New Delhi
Singapore Sydney Toronto

1 2 3 4 5 6 7 8 9 0 DOC/DOC 1 2 0 9 8 7 6 5 4

ISBN 978-0-07-182283-1
MHID 0-07-182283-6

The pages within this book were printed on acid-free paper.

Sponsoring Editor
Judy Bass

Editorial Supervisor
Stephen M. Smith

Production Supervisor
Pamela A. Pelton

Acquisitions Coordinator
Amy Stonebraker

Project Manager
Patricia Wallenburg, TypeWriting

Copy Editor
James K. Madru

Proofreader
Claire Splan

Indexer
Judy Davis

Art Director, Cover
Jeff Weeks

Composition
TypeWriting

McGraw-Hill Education books are available at special quantity discounts to use as premiums and sales promotions or for use in corporate training programs. To contact a representative, please visit the Contact Us page at www.mhprofessional.com.

CONTENTS

PREFACE

Finding the *Invisible* Low-Hanging Fruit

When I first got into quality improvement in 1990, I frequently heard consultants speak about "low-hanging fruit" just waiting to be picked. Two years later and thousands of staff hours later, I still hadn't found any low-hanging fruit. Over the last two decades, I have heard from hundreds of customers whose companies implemented Six Sigma only to have similar dismal results.

In any company, if there really is low-hanging fruit, it's usually *visible* everywhere from the factory floor or nursing unit to the management conference room. When it's that visible, anyone can pick it with a little common sense and a bit of trial and error.

This is why there is no *visible* low-hanging fruit. Somebody has already picked it! And this is what stops most leaders from even considering the tools of breakthrough improvement—they can't see any more fruit to be picked. And, because they can't see it, they come to believe that *their problems are unsolvable*, so they stop looking.

In company after company, though, my own included, I have found orchards filled with *invisible* low-hanging fruit. You just can't see it with the naked eye.

You can, however, discern it through the magnifying lens of Excel PivotTables and charts. They make the seemingly invisible visible. They are the microscope, the MRI, the EKG of business diagnosis.

When Louis Pasteur said that there were tiny bugs in the air and water, everyone thought he was crazy because they weren't visible . . . to the naked eye. Everyone thought it was just an "ill wind" that made people sick.

In today's tough economic times, everyone laments about how hard it is, how an "ill wind" has blown through their business, their industry, their economy. But have they considered using the modern tools of business medicine to root out the infectious agents in their business? Have they taken the time to look for the *invisible* low-hanging fruit in their businesses? I doubt it.

Someone sent me an e-mail the other day that said that even in the poorest run companies, he'd had no luck finding the low-hanging fruit. But in every company I've ever worked with, I've found millions of dollars just waiting to be retrieved from the orchards of delay, defects, and deviation—the three silent killers of productivity and profitability.

Are you looking for the obvious? Or investigating the invisible?

The low-hanging fruit is always invisible to the naked eye. Turn the magnifying and illuminating tools of the QI Macros and Excel on your most difficult operational problems and stare into the depths of the unknown, the unfamiliar. You'll invariably find bushels of bucks, just waiting for a vigilant harvester.

Every Picture Tells a Story

I remember when Rod Stewart released the album *Every Picture Tells a Story*. During a recent training, I was surprised to rediscover that this embodies the message of *breakthrough improvement*, and I was surprised by how many participants had no idea that your data needs to tell a story. One of the biggest "ah-has" I got out of this training was that your data has to lead you to what needs improvement and what can be left alone.

The secret of all data-based improvement methods is to become a crime scene investigator: let the facts paint a picture and lead you step by step to the cause of your problems with speed, quality, or costs. You can't let personal bias enter into the detective work. You let the forensic evidence lead you to the root cause and solution. Like a crime scene where there are hundreds of fingerprints and distracting bits of hair and fiber, most of which has nothing to do with the crime, *most of your data is a distraction, not evidence*. To be a great data sleuth, you have to learn to discard huge amounts of information, focusing on the few tidbits that lead you to the root cause. Then you transform those bits of data into control charts, Pareto charts, histograms, and fishbone diagrams to get every picture to tell a story.

Most companies are choking on the amount of data they collect. Most of this is transformed into a series of mind-numbingly identical graphs that hide the bad and accentuate the good so that everyone can go on feeling good about their performance. These pictures rarely tell a story.

The Goal

The goal is to learn how to use the QI Macros and Excel to build a business case for improvement that workers, managers, and leaders can understand quickly and easily. To do so means moving beyond the old problem-solving methods of common sense, trial and error, and gut feel. We have to learn to quantify common sense and turn it into actionable intelligence. This book will move you from hopeful problem solving to competent data analysis and breakthrough improvement (Fig. P.1).

I have found over the years that you don't need to know everything about Excel to start finding the low-hanging fruit. The Magnificent Seven Tools (see Chap. 4) will solve 99 percent of the problems facing any business. And there's a step-by-step method for using these tools to create compelling business cases that anyone can understand. Moreover, if you really want to solve the remaining 1 percent, you can go on to learn all the tools of Lean Six Sigma, but they aren't necessary to start making breakthrough improvements in speed, quality, productivity, and profits.

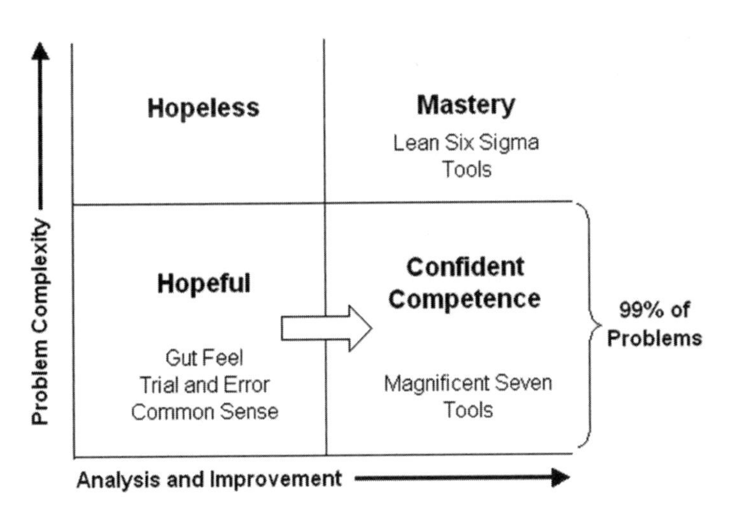

Figure P.1 Confident competence.

As much as a third of a company's total revenue is spent fixing problems that shouldn't exist (low-hanging fruit). As you learn how to use the Magnificent Seven Tools of the QI Macros and Excel, you'll reap a bountiful harvest that will catapult your business ahead of the competition. I've worked with teams that found million-dollar savings in an afternoon. Isn't it time to start using the mountains of data in your business to start plugging the leaks in your cash flow? It's easy, and I'll show you how.

Problems You Can Solve with QI Macros and Excel

Some people think that problem solving with Excel is restricted to financial data. Not true. With Excel you can solve the three key problems facing any business: process problems, profit problems, and people problems.

Process Problems

- Accounts payable
- Accounts receivable
- Billing/invoicing
- Call centers
- Customer service
- Delivery
- Distribution
- Engineering
- Facility management
- Finance
- Information systems
- Innovation
- Inventory
- Manufacturing
- Marketing
- Operations
- Payroll
- Product development
- Research and development
- Sales
- Service

Profit Problems

- Sell more
- Sell more often
- Sell at full price
- Avoid discounts
- Open new markets
- Expand your product line
- Cross-sell
- Up-sell
- Open new channels
- Raise prices
- Boost your margins
- Reduce unit cost

- Cut costs
- Be more efficient
- Speed up time to market
- Franchise
- Conserve cash

People Problems

- Recruiting
- Retaining
- Employee engagement
- Recognition and reward
- Staff utilization
- Teamwork
- Communication
- Collaboration
- Turf wars
- Negativity
- Complacency

Solving process problems will solve many profit and people problems. When everyone becomes engaged in serving customers in better, more useful ways and using data to drive decisions, conflict tends to dissolve and profits rise.

Good Data, Bad Decisions

In the April 2012 issue of the *Harvard Business Review*, Shvetank Shah, Andrew Horn, and Jaime Capella argue that "Good Data Won't Guarantee Good Decisions." I find that many improvement teams get stuck on the idea of needing *good data*. They often get lost trying to collect *better* data. Unfortunately, even good data won't guarantee a good decision. The authors note: "At this very moment, there's an odds-on chance that someone in your organization is making a poor decision on the basis of information that was enormously expensive to collect." Ouch!

The authors found that there are three main types of decision makers:

- *Visceral* (i.e., gut feel) decision makers
- *Unquestioning empiricists*, who rely on the numbers alone
- *Informed skeptics*, who combine their gut with data

The authors also say that "too many companies are stuck in the 'expert' phase," where only a few people know what to do with performance data. The knowledge of how to analyze the data hasn't spread to the masses.

Good Data Is Hard to Find

The authors also found that fewer than half of employees know where to find the data they need to do analysis. Some improvement teams get stuck waiting for information technology to get the data into a central repository rather than collecting a little data manually to move their project forward.

Get the Right Tools

The authors also say that "having the right tools to create and interpret data displays is vital. Half of all employees find that information is in an unusable format. The best companies provide charts instead of raw data." Spreadsheets are unusable by most employees, but with the QI Macros you can quickly create charts of data in those spreadsheets that everyone can understand.

Download a 90-day trial of the QI Macros for Excel at www.qimacros.com/breakthrough. In addition, you will find links to YouTube videos for each chapter of this book at www.breakthrough-improvement-excel.com. To find out more about any topic in this book, Google "topic site: qimacros.com," or you can use the search function on most pages at www.qimacros.com.

Special Offer: Save more than the cost of this book when you purchase the QI Macros. Go to www.qimacros.com/biqixl to order now.

While there are endless ways to use Excel and the QI Macros, this book will focus on just the key tools that have the greatest impact. Master the Magnificent Seven Tools, and they will take you a long way toward maximizing productivity and profitability.

If you want to know how to set up your data for breakthrough improvement, start with Chapter 1. If you already know formats, formulas and filtering, start with Chapter 4. If you want to jump right into the heart of breakthrough improvement, go to Chapter 9.

Let's start finding the invisible low-hanging fruit in your company using the QI Macros and Excel.

Jay Arthur

Breakthrough Improvement with QI Macros and Excel®

CHAPTER 1

Spreadsheet Design for Breakthrough Improvement

People sometimes make their data hard to analyze or chart by setting it up incorrectly. All too often people try to make their spreadsheet *easy to read.* After 20 years of working with Excel, I can tell you that there are only two kinds of people that can actually read a spreadsheet: accountants and finance managers. And even those folks prefer to see the data as a chart. Therefore, instead of trying to make the spreadsheet easy to read, *make it easy to analyze and chart.* Ease and difficulty are determined by

- Horizontal versus vertical layout
- Headings
- Formatting
- Summarizing

Horizontal or Vertical?

Because most calendars show time horizontally, not vertically, most people tend to organize their data by dates and times horizontally. And sometimes they stack rows of data (e.g., a 4 × 3 table of monthly data) that cannot be charted easily with Excel. The QI Macros will draw a chart of horizontal or vertical data, but the best way is vertical. Why? Because there's a lot more rows than columns available (Fig. 1.1). In earlier versions of Excel, you were limited to 255 columns horizontally (so there's no way to input 365 values for each day of the year). Newer versions of Excel don't have this restriction, but you can still see more rows than columns on a typical screen, and it makes it easier to print.

	A	B	C	D	E	F	G	H	I	J	K	L	M	N
1	Subgroup	1	2	3	4	5	6	7	8	9	10	11	12	13
2	x1	5.4	4.8	7.2	4.7	4.5	5.7	4.9	6.0	5.9	3.5	3.9	5.6	3.8
3	x2	6.2	6.3	4.6	4.3	5.7	4.2	3.7	3.0	3.9	5.7	4.2	4.5	3.2
4	x3	5.5	4.0	5.6	6.9	5.3	2.7	4.3	5.3	6.5	3.2	6.2	5.4	4.9
5	x4	5.1	6.0	5.1	6.8	5.4	5.1	4.1	3.9	2.7	3.6	6.3	4.4	5.0
6	x5	4.8	4.1	6.6	4.4	5.3	5.9	5.4	5.3	4.6	5.2	4.0	6.6	6.4
7								Horizontal						
8	Subgroup	x1	x2	x3	x4	x5		255 Columns						
9	1	5.4	6.2	5.5	5.1	4.8								
10	2	4.8	6.3	4.0	6.0	4.1								
11	3	7.2	4.6	5.6	5.1	6.6								
12	4	4.7	4.3	6.9	6.8	4.4								
13	5	4.5	5.7	5.3	5.4	5.3								
14	6	5.7	4.2	2.7	5.1	5.9		Vertical						
15	7	4.9	3.7	4.3	4.1	5.4		65,535 Rows						
16	8	6.0	3.0	5.3	3.9	5.3								
17	9	5.9	3.9	6.5	2.7	4.6								
18	10	3.5	5.7	3.2	3.6	5.2								
19	11	3.9	4.2	6.2	6.3	4.0								

Figure 1.1 Horizontal versus vertical data.

TIP Given a choice, start with vertical data.

Column and Row Headings

Having settled on vertical, it's time to think about headings. Every column and row should have a heading (Fig. 1.2).

	A	B
1		Column Heading
2	Row Heading 1	
3	Row Heading 2	
4	Row Heading 3	
5	Row Heading 4	
6	Row Heading 5	

Figure 1.2 Row and column headings.

Column Headings

The QI Macros will use column headings for chart titles and axis labels, so make the title descriptive.

Row Headings

If the data is organized by time, then the row headings should be

- Dates (see Fig. 1.4)
- Times
- Batch or subgroup (see Fig. 1.1) (If this is a number such as 1, 2, or 3, consider using an apostrophe in front of the number to force it to be text. In this way, the QI Macros and Excel cannot be confused about whether it's data or a label.)
- Descriptive labels [If the data is a category such as *type of defect*, then use *descriptive labels* (Fig. 1.3) for each category instead of abbreviations or codes (e.g., "FF" for folded flaps or "BF" for bent/damaged flaps).]

TIP Use color or shading for visual clarity.

If your data has a numerator and denominator (e.g., errors per sample), put them in adjacent columns with a leading heading (Fig. 1.4).

	A	B	C	D	E
1	**Carton Manufacturing Defects**				
2		**Line 1**	**Line 2**	**Line 3**	**Total Defects**
3	Folded flaps	16	6	83	105
4	Bent/Damaged flaps	37	22	24	83
5	Carton will not open	29	18	29	76
6	Poor ink adhesion	7	8	18	33
7	Off color	14	5	12	31
8	Ink smears/streaks		5	19	24
9	Oil spots			14	14
10	Fisheye	9			9
11	Missing color			8	8
12	Mislabeled			3	3
13	Damaged Pallet	3			3
14	Undercount		2		2

Figure 1.3 Category row headings.

	A	B	C
1		Discrepancies	Sample (n)
2	Jan-19	8	968
3	Jan-20	13	1216
4	Jan-21	13	804
5	Jan-24	16	1401
6	Jan-25	14	1376

Figure 1.4 Numerator-denominator headings.

Common Heading Mistakes

The most common and troubling mistakes include using multiple rows for headings (Fig. 1.5). This makes it impossible to easily chart the data. Excel treats "Line" as a title and 1-2-3 as numbers for the chart.

Instead, put "Line 1" in a single cell, and right-click on the cell to use Format Cells "Wrap text" (Fig. 1.6) to fit the text into the cell.

	A	B	C	D	E
1	Type of	Line	Line	Line	Total
2	Defect	1	2	3	Defects

Figure 1.5 Multiline headings.

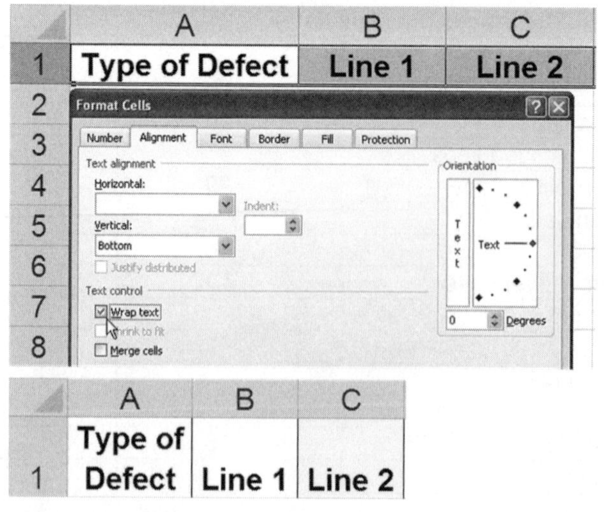

Figure 1.6 Wrapping title text.

Formatting Mistakes

Don't merge cells such as "Circuit Board Defects" (Fig. 1.7). Merged cells are particularly hard to use when copying, pasting, or charting data.

TIP Instead of "Merge Cells," use "Center Across Selection" (Fig. 1.8) to accomplish the same visual result without hog-tying copy/paste/chart capabilities.

	A	B
1	Circuit Board Defects	
2	Bad	8
3	Bridge	1
4	Damage	6
5	Excess	3
6	Extra	1

Figure 1.7 Don't merge cells.

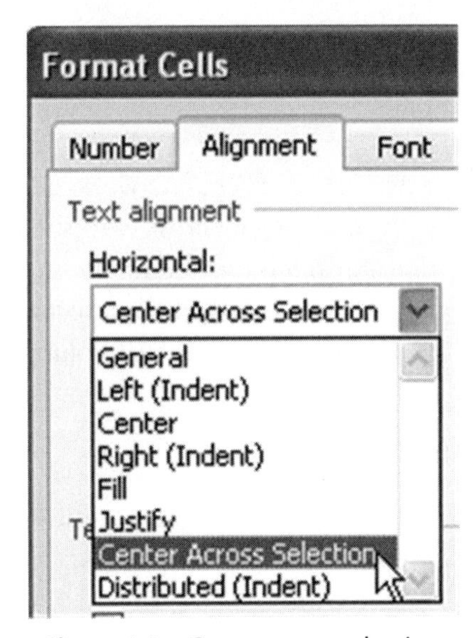

Figure 1.8 Center across selection.

	A	B	C	D
1	Month	Falls/1000 Patient Days (2004)	Falls/1000 Patient Days	Falls/1000 Patient Days (2006)
2	Jan	3.6	7.5	4.3
3	Feb	4.5	7.2	3.9
4	Mar	4.7	4.6	3.9
5	Apr	6.0	3.6	4.5
6	May	4.6	8.2	4.0
7	Jun	3.6	9.6	
8	Jul	7.6	6.0	
9	Aug	7.7	3.8	
10	Sep	5.6	5.4	
11	Oct	5.7	6.6	
12	Nov	7.0	4.8	
13	Dec	5.3	4.4	

Figure 1.9 Don't split data into columns.

Don't put your time-series data side by side (Fig. 1.9) because you will have to stack it to draw a line, run, or control chart.

It's okay, however, to show category data side by side (see Fig. 1.3) because it's a good way to make comparisons.

Data Collection for Breakthrough Improvement

First, *put all your data on one worksheet.* I've seen annual data split into quarters, with each quarter put on a separate worksheet. You can't use the QI Macros or Excel if your data is split across worksheets.

Second, *don't put quarterly totals and grand totals in line with your data* (Fig. 1.10, columns A and B) because you have to select around them to draw a chart. Put them beside your data (see Fig. 1.10, columns D through K).

Dealing with Difficult People

I know that some manager will ask to see it in a different, more complex way, but keep your raw data clean and simple. Use the QI Macros "Paste Link" or "Paste Link Transpose" functions to make a linked copy of the data that you can modify to meet any need while keeping the raw data clean.

	A	B	C	D	E	F	G	H	I	J	K
1	Month	Falls/1000 Patient Days		Month	Falls/1000 Patient Days			Quarterly Average			Annual Average
2	Jan-04	3.6		Jan-04	3.6		2004 Q1	4.3		2004	5.5
3	Feb-04	4.5		Feb-04	4.5		2004 Q2	4.8		2005	6.0
4	Mar-04	4.7		Mar-04	4.7		2004 Q3	7.0		2006	4.1
5	Q1 Ave	4.3		Apr-04	6.0		2004 Q4	6.0			
6	Apr-04	6.0		May-04	4.6		2005 Q1	6.4			
7	May-04	4.6		Jun-04	3.6		2005 Q2	7.1			
8	Jun-04	3.6		Jul-04	7.6		2005 Q3	5.1			
9	Q2 Ave	4.8		Aug-04	7.7		2005 Q4	5.3			
10	Jul-04	7.6		Sep-04	5.6		2006 Q1	4.1			
11	Aug-04	7.7		Oct-04	5.7		2006 Q2	4.2			
12	Sep-04	5.6		Nov-04	7.0						
13	Q3 Ave	7.0		Dec-04	5.3						

Figure 1.10 Don't insert totals in data.

If, for example, you take the data in Figure 1.10 and use the QI Macros "Paste Link Transpose" function, you will get data that has links back to the original data (Fig. 1.11). You can insert quarterly averages for reporting purposes but still have access to the original data for charting.

	A	B	C	D	E	F	G
1	Month	Jan-04	Feb-04	Mar-04	Apr-04	May-04	Jun-04
2	Falls/1000 Patient Days	3.6	4.5	4.7	6.0	4.6	3.6

Figure 1.11 Transpose data if needed.

Defect Tracking with Microsoft Excel

Name any sport—baseball, soccer, football, or basketball—and there's always a method of keeping score. And sports teams don't just track wins; they also track mistakes and errors. Baseball tracks all kinds of statistics: runs, hits, walks, errors, and batting average, to name a few. There's even a book, *MoneyBall*, by Michael Lewis that tells the story of how the Oakland A's put together a winning team for an affordable amount of money using these statistics. Companies can do the same thing.

And most companies do. They collect numbers about everything but rarely examine what they've collected. Therefore, the first step in breakthrough improvement is to look at what's already being collected and examine it to find the invisible low-hanging fruit.

TIP Start with data that already exists; it's faster.

It's easy for people to detect mistakes and errors when they happen often enough. But when the frequency of those mistakes falls below a certain level or they have a minor impact, you can no longer detect them with your five senses. You need some better tools. Fortunately, the process and tools are simple.

1. Count your misses, mistakes, and errors (PivotTable® and control chart).
2. Categorize your misses, mistakes, and errors (Pareto chart).
3. Fix the biggest problem first (worst first!).

How to Set Up a Defect-Tracking Worksheet

First, figure out what you want to track, and make a series of headings. To keep a log of defects (Fig. 1.12) that can be used in a PivotTable, you need a minimum of:

1. *A date or time.* When did it happen?
2. *Location.* Where did it happen?
3. *Type.* What was the defect, mistake, or error?
4. *Cause.* Why did the error occur?
5. *Description.* Free-form description that can be analyzed with the QI Macros "Word Count."
6. *Count of number of times it occurred.*

Put these labels in row 11 (we'll use rows 1 through 10 to set up correct data values) (Fig. 1.12).

Next, you need to make sure that people enter the defect data consistently. To do this, use Excel's "Auto Fill" and "Data Validation" functions. Enter values that you want Excel to autocomplete in cells B1 through D10 (Fig. 1.13). Adjust the values to meet your needs. For Excel to complete a value when someone types in a cell, the values have to be stacked above the heading. No blanks.

11	Date	Location	Defect - Mistake - Error Type	Cause	Description	Count

Figure 1.12 Log of defects.

	A	B	C	D	E	F
1	1/1/2013					
2	12/31/2020					
3						
4			Bent/Damaged Flaps			
5			Carton will not open			
6			Folded Flaps			
7			Ink smears/stains			
8		Line 1	Poor Ink Adhesion	Machine Drift		
9		Line 2	Off Color	Setup		
10		Line 3	Fisheye	Excess Glue		
11	**Date**	**Location**	**Defect - Mistake - Error Type**	**Cause**	**Description**	**Count**

Figure 1.13 Excel auto fill and data validation.

- Cells A1 and A2 contain the valid start/end dates or times for dates in column A.
- Rows 1 through 10 have lists of valid locations and types of defects.

Adding items above the labels allows Excel's "Autofill" function to simplify data entry. Don't allow an "Other" category; it's useless for analysis. Make users enter something useful.

Next, select column A, and choose "Data-Data Validation" (Fig. 1.14). Then set the validation criteria to be "Date" with a start and end date (cells A1 and A2).

Then, for columns B through D, select the cell range and choose "Data-Data Validation" (Fig. 1.15), but this time choose "List" and specify cells 1

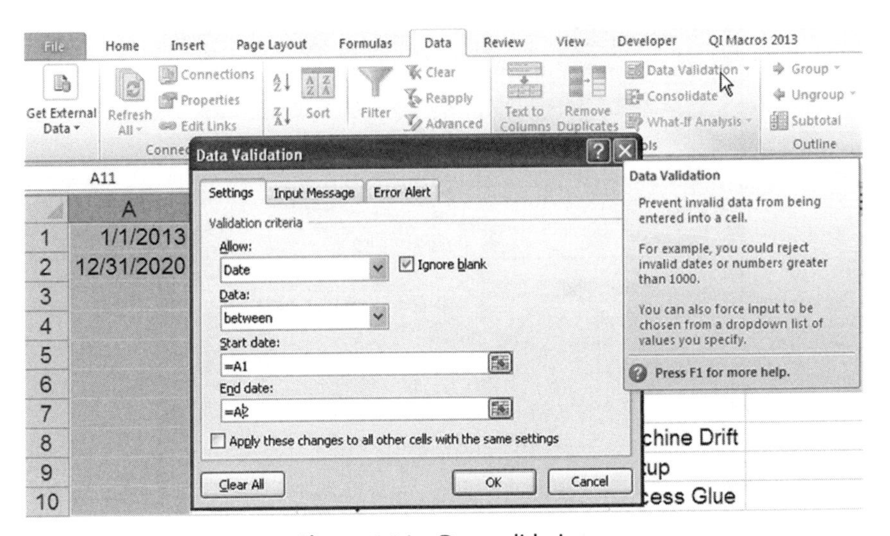

Figure 1.14 Date validation.

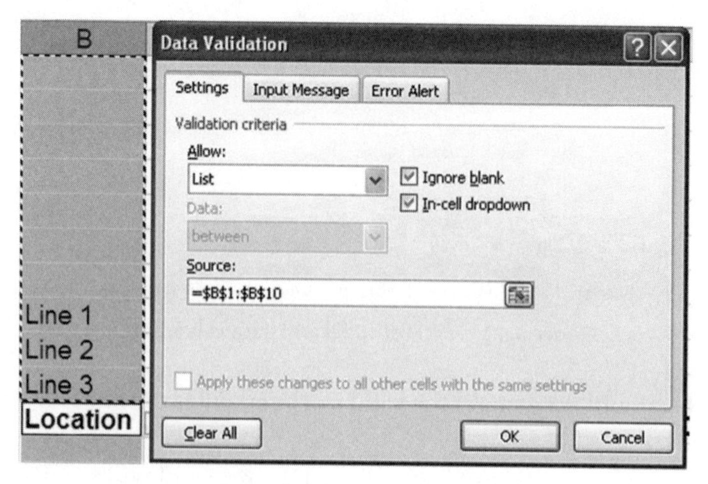

Figure 1.15 List validation.

through 10 above the heading. This will give users a choice of valid inputs, which will make it easier to analyze with Excel PivotTables.

Then specify an error message if they enter something incorrect (Fig. 1.16). You can choose "Stop" or just a warning. A warning will allow them to add new criteria in rows 1 through 10, which will be useful as new criteria arise (Fig. 1.16).

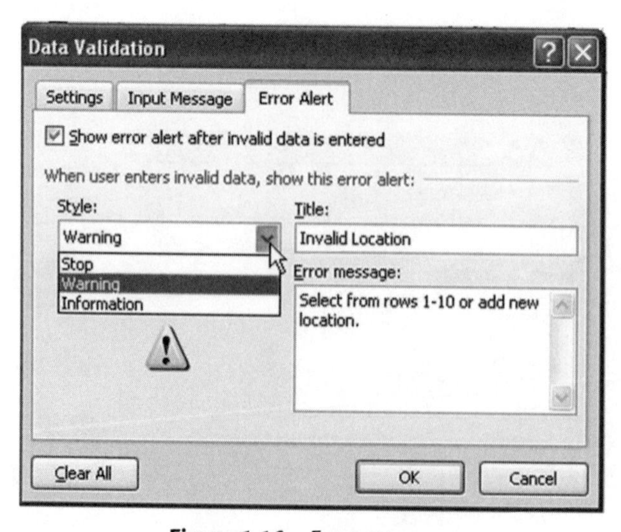

Figure 1.16 Error message.

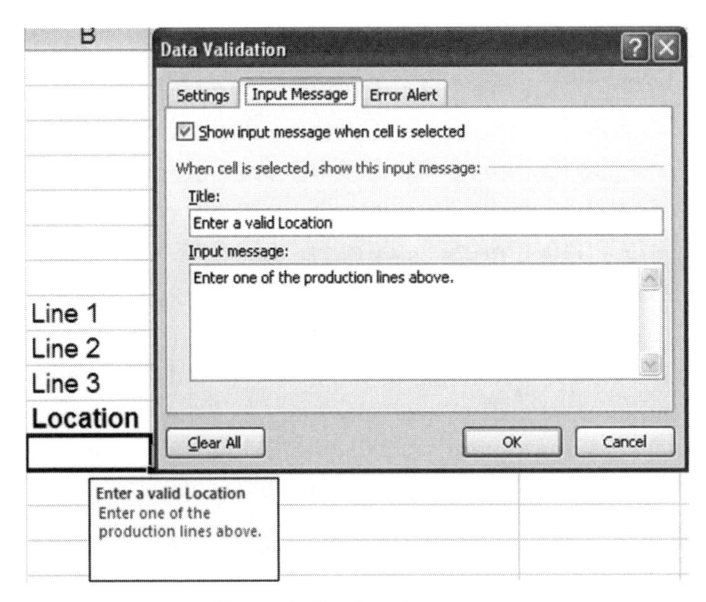

Figure 1.17 Input message.

Then select the first cell in row 12 and enter an input message (Fig. 1.17). This will give new users some starting help. It may not be useful on every cell in the column.

Now, when data is entered, it will be entered correctly, and criteria can be updated as needed. Users have a choice of clicking on the arrow next to a cell to select a value or simply typing and letting Excel's "Auto Fill" work its magic (Fig. 1.18).

This will ensure that data is entered and formatted correctly (Fig. 1.19). Or there may be defect codes or detailed descriptions of how the error occurred or what caused it. These can be included in the spreadsheet for analysis.

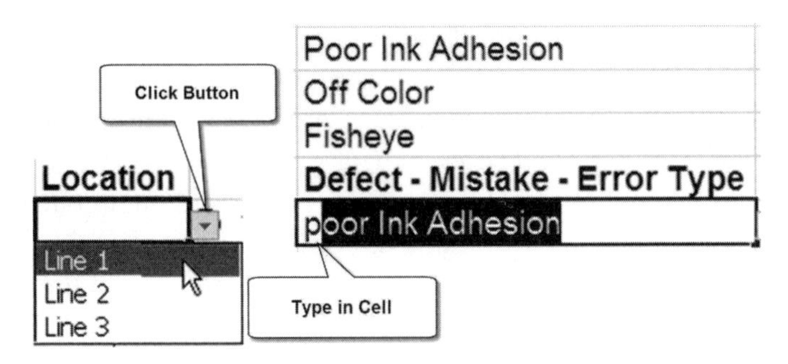

Figure 1.18 Pulldown menu or auto fill data entry.

11	Date	Location	Defect - Mistake - Error Type
12	1/2/2010	Line 1	Bent/Damaged flaps
13	1/2/2010	Line 1	Carton will not open
14	1/2/2010	Line 2	Bent/Damaged flaps
15	1/2/2010	Line 2	Carton will not open
16	1/2/2010	Line 3	Folded flaps

Figure 1.19 Defects with counts.

This is a simple system for tracking defects and readying them for analysis and improvement-project selection.

Rules for creating a spreadsheet that will help Excel and the QI Macros make intelligent choices for you include the following:

1. *Use a single row of headings.* Don't put one word in row 1 and the next word in row 2.
2. *Format cells to "word-wrap" headings in a single cell, and format them as bold.*
3. *Don't repeat headings.* Use a unique heading for each column.
4. *No blank rows or columns.* Excel thinks a blank row or column means the end of the data. Always use a unique heading so that the data can be used with PivotTables, sort, filter, and other tools.

While most companies do some tracking of their defects, mistakes, and errors, the data is usually too inconsistent for easy analysis. *Solution:* Use Excel to mistake-proof defect tracking.

Warning: Mistake-Proof Data Collection

Chip and Dan Heath, authors of *Decisive*, say, "People [are] collecting the data, and they don't realize they're cooking the books." One of the problems with data collection is *confirmation bias—people try to use data to confirm what they already think and believe.*

When I worked in the phone company, I worked on an out-of-state improvement project for two months. Lead times to repair a customer's service were running four days. Unfortunately, the leaders of the "repair" part of the company were convinced that they "needed more repair technicians." So, along with the quality consultants they'd hired, they chased data that supported their thoughts and beliefs.

About one day into the project, I started trying to convince the leadership that they didn't need more people—*they needed less repair work!* Needless to say, that idea didn't go over very well. When 51 percent of the calls to a company are for repair, the focus shouldn't be on fixing things faster; it should be on eliminating the need to fix things. After two months trying to force-fit the data to their solution, the project ended in failure.

Lesson: Don't let your desired solution lead you to data to support it; instead, let the data lead you to the solution. It's a lot faster and creates better results.

These are some simple ways to make your data life easy or hard. It's up to you to implement them.

Data for Service Companies: Defects and Turnaround Times

Service companies are concerned with two things: defects (i.e., mistakes, errors, omissions, etc.) and turnaround times. Therefore, for most defect-related applications, we need a date, the number of defects, the sample size (total number of opportunities for a defect), and the percentage of nonconforming units. These four columns give you the ability to create the most common control charts in service industries (see Chap. 10). Figure 1.20 shows what the column headings would look like in Microsoft Excel.

	A	B	C	D
1	Sample Number	Defects	Sample Size	Percent Defects
2	S1	12	100	12%
3	S2	8	80	10%
4	S3	6	80	8%
5	S4	9	100	9%

Figure 1.20 Defect data.

It's important to use a reasonable time interval such as hours, days, weeks, or months. Hospitals often report by quarter, but you would need five years of data (20 quarters) to get enough data to draw a useful control chart. Two years of monthly data (24 months) would give a useful control chart but may not be granular enough for some applications.

One of my customers asked me, "I have a back room operation; how do I track the number of times people don't follow established procedures?"

"Simple," I replied. "Just adapt the columns to your data." The result is shown in Figure 1.21.

In health care, hospitals track hand washing because hand washing is the key to preventing hospital-acquired infections that, according to the Centers for Disease Control and Prevention, kill 99,000 people a year. Figure 1.22 shows how the data would appear.

This format also can be used for defect data concerning patient falls, infection rates, and other hospital events (Fig. 1.23).

Once you understand the nonconforming/sample/ratio format for defect data, it's easy to lay out data for charting. Using the QI Macros, it's

	A	B	C	D
1	Sample Number	Procedures Not Followed	Total Opportunities	Percent Not Followed
2	1-Mar	12	100	12%
3	2-Mar	8	80	10%
4	3-Mar	6	80	8%
5	4-Mar	9	100	9%

Figure 1.21 Adapted defect tracking worksheet.

	A	B	C	D
1	Sample Number	Hands Not Washed	Total Opportunities	Percent Not Washed
2	1-Mar	12	100	12%
3	2-Mar	8	80	10%
4	3-Mar	6	80	8%
5	4-Mar	9	100	9%

Figure 1.22 Healthcare hand washing tracking worksheet.

	A	B	C	D
1	Month	Total Patient Falls	Total Patient Days	Falls/1000 Patient Days
2	Jul	17	4658	3.6
3	Aug	22	4909	4.5
4	Sep	23	4886	4.7
5	Oct	30	4970	6.0

Figure 1.23 Patient falls per 1,000 patient days.

easy to create *p*, *u*, or XmR charts from this data. Use nonconforming/ sample size (columns A through C) for *p* or *u* charts (Fig. 1.24); use the ratio (columns A and D) for XmR charts (Fig. 1.25).

TIP You don't have to know what chart to choose. Just select the data, and click on the QI Macros "Control Chart Wizard." It will figure out what chart to draw based on the data selected.

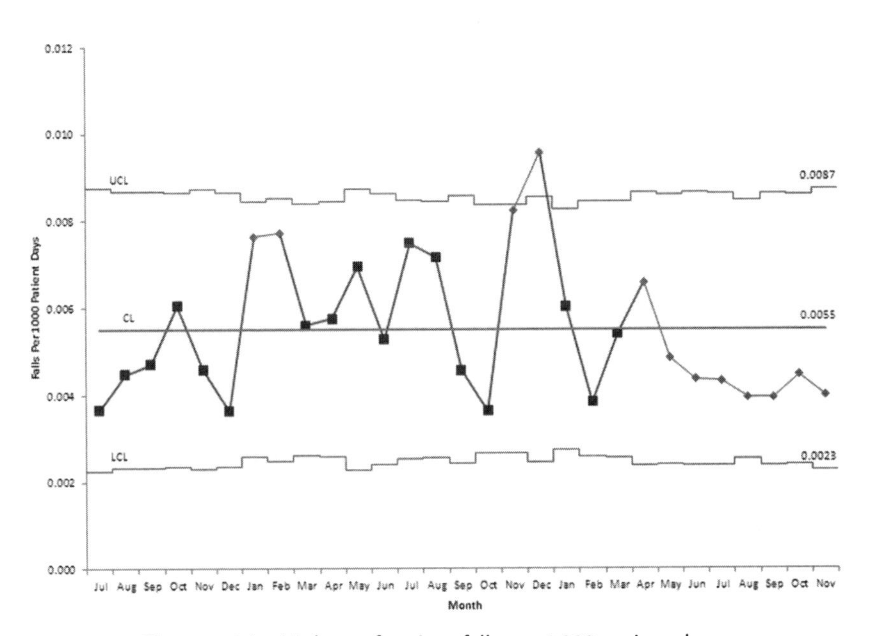

Figure 1.24 U chart of patient falls per 1,000 patient days.

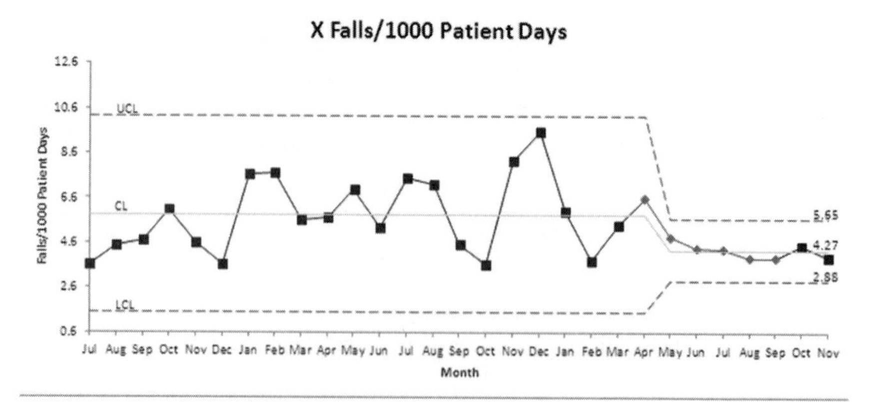

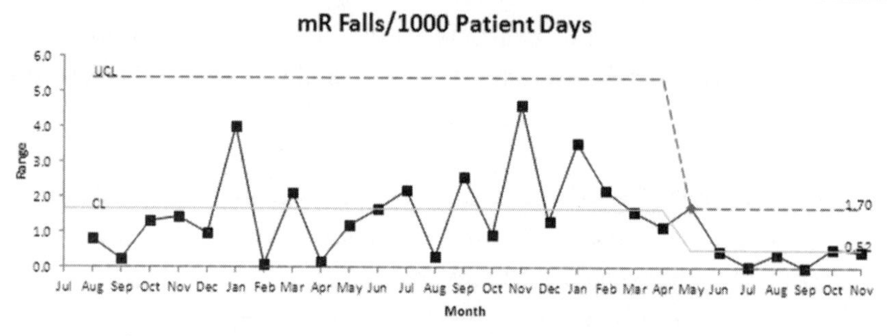

Figure 1.25 XmR chart of patient falls per 1,000 patient days.

I often provide both types of charts because the variable limits of a *p* or *u* chart sometimes confuse viewers. Having to explain why the limits vary sometimes detracts from what the data is telling us.

Time Data

From a turnaround-time perspective, service industries often track response times on a pass/fail (i.e., defect) basis: number of calls answered in 60 seconds (call center), number of cardiac patients given aspirin at arrival (emergency room), number of quotes delivered within 24 hours (insurance), and so on. This takes on the form of having met or missed a service commitment, as shown in Figure 1.26.

While most managers like to know how well they are doing (percent of commitments met), breakthrough improvement problem-solving methods work best when focused on the problem (percent of commitments missed). Figure 1.27 shows managerial performance data.

	A	B	C	D
1	Sample Number	Missed Commitments	Total Commitments	Percent Missed
2	1-Mar	12	100	12%
3	2-Mar	8	80	10%
4	3-Mar	6	80	8%
5	4-Mar	9	100	9%

Figure 1.26 Commitments missed.

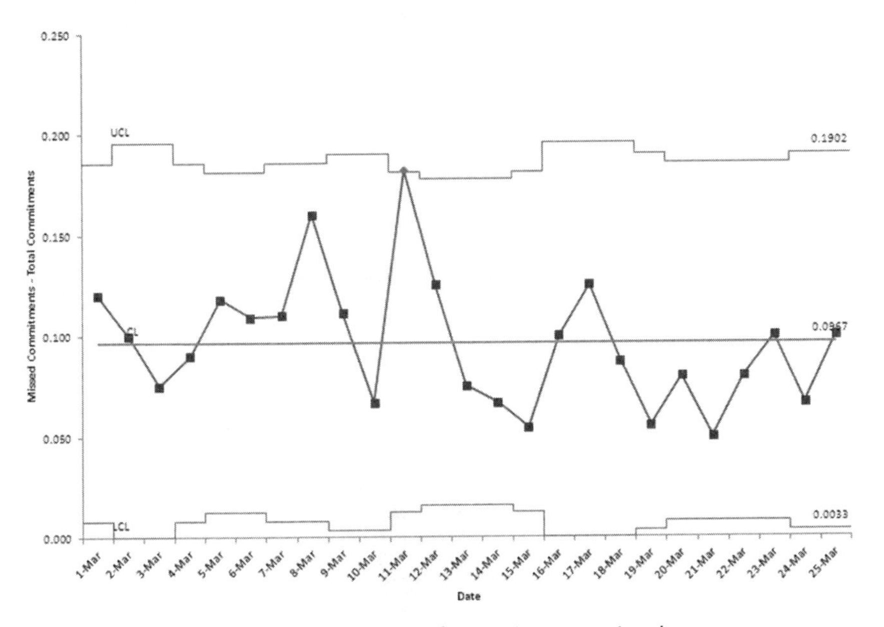

Figure 1.27 P chart of commitments missed.

Turnaround-time data also can be shown in actual days, hours, or minutes. This could cover turnaround times in a hospital or days to process an insurance claim. The data can be easily charted as an XmR chart (Fig. 1.28) or a histogram (Fig. 1.29) using the QI Macros.

In this example, we can see that the process for paying claims is predictable, but we might incur additional charges for exceeding 30 or 45 days before claims are paid depending on the supplier's requirements.

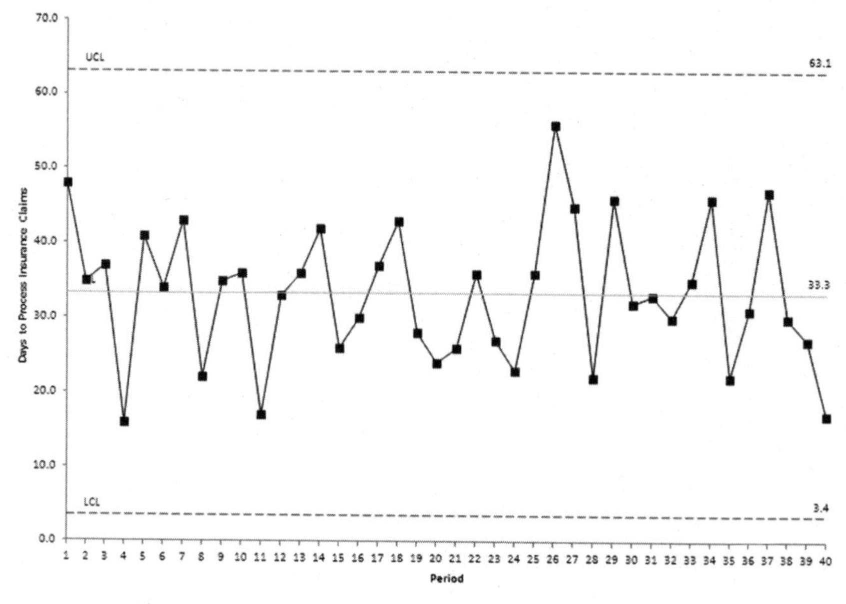

Figure 1.28 X chart of days to process insurance claims.

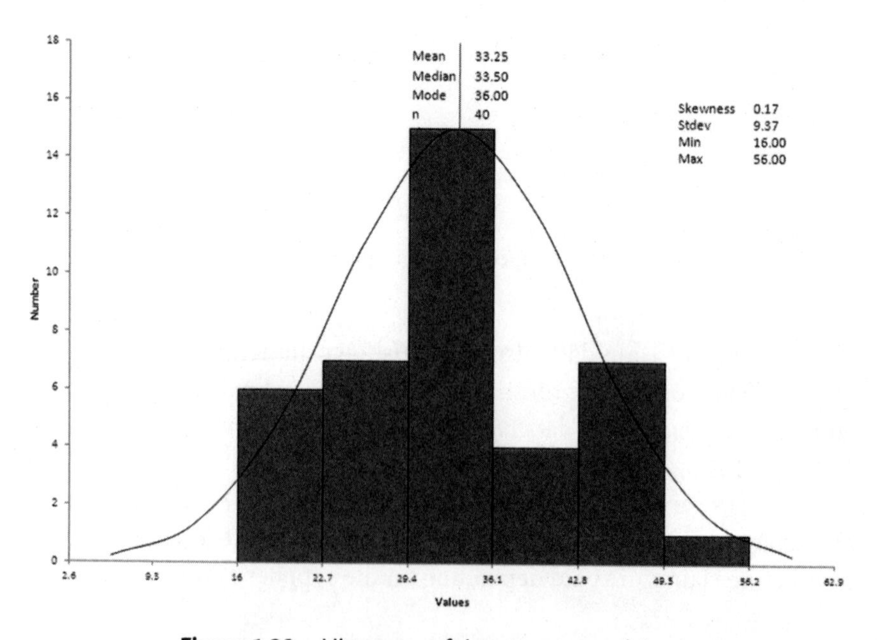

Figure 1.29 Histogram of time to process claims.

Summary

Breakthrough improvement begins by getting the data in the right format for Excel and the QI Macros. We can then use Excel's PivotTables on defect-tracking worksheets to summarize the data and ready it for the QI Macro control charts, Pareto charts, or histograms.

We can organize time-series data into columns that can be easily charted. Once the data is organized into columns, it's easy to turn the data into a control chart. The most common control charts used in service industries are the *p*, *u*, and XmR charts. They can be used to monitor error rates, missed commitments, and turnaround times. Just use these simple formats (see Fig. 1.20) as a guide to start collecting data in Excel.

CHAPTER 2

Excel Formulas for Breakthrough Improvement

Some users of the QI Macros miss the opportunity to use Excel's formulas to prepare data for charting or analysis. You'll find examples of Excel formulas in the QI Macros "Improvement Tools–Excel Formulas." Here are some ways to use Excel to your advantage.

Calculate Ratios

Most companies need to analyze defects in relationship to overall volume (e.g., repeat calls to a call center, repeat visits to an emergency room for the same problem, returned product, etc.). Although *p* and *u* control charts will show this kind of data, it's also acceptable to turn the numbers into a ratio and use an XmR (individuals and moving range) chart.

Most hospitals keep track of patient falls, and they have to report it as falls per thousand patient days (Fig. 2.1). So how do you express it as a ratio?

Simply click on the cell to right of the data, and enter the formula starting with an equal sign:

=A3/(B3/1000) or = (A3*1000)/B3

	A	B	C	D
1	Total Patient Falls	Total Patient Days	Convert B3 to 1000 Pt Days	Falls/1000 Patient Days
3	17	4658	4.658	=(A3*1000)/B3

Figure 2.1 Excel formula for falls per 1,000 patient days.

where

- A3 is number of falls.
- B3 is number of patient days.

Putting formulas in parentheses tells Excel to do that operation first and then anything else; otherwise, the answer could be incorrect. A3/B3/1000, for example, would be incorrect because Excel would do it in the order shown.

Then simply move the mouse over the lower-right corner of the cell containing the formula (look for the cursor to change to a plus shape), and double-click (Fig. 2.2). Excel will copy the formula down to the last nonblank cell.

Then just select the ratios in column D to draw a control chart using the QI Macros "Control Chart Wizard." Why do we need a control chart? Because it will tell us if the process is stable and predictable or erratic.

	A	B	C	D
1	Total Patient Falls	Total Patient Days	Convert B3 to 1000 Pt Days	Falls/1000 Patient Days
3	17	4658	4.658	3.6
4	22	4909	4.909	4.5
5	23	4886	4.886	4.7
6	30	4970	4.97	6.0

Figure 2.2 Copying formulas.

Excel Will Guide You

If you don't know what function to use, Excel can show you the way. Simply select an empty cell, and click on "Insert Function f_x." Excel will show you the functions available (Fig. 2.3).

Text Formulas

I have found that people often use a single cell to store more than one kind of data. Sometimes they combine text with numbers (e.g., "Debit$21.95" or "[Tools]Screwdriver") without any useful structure or delimiters. Sometimes the only way to get at this data is to use formulas to extract it.

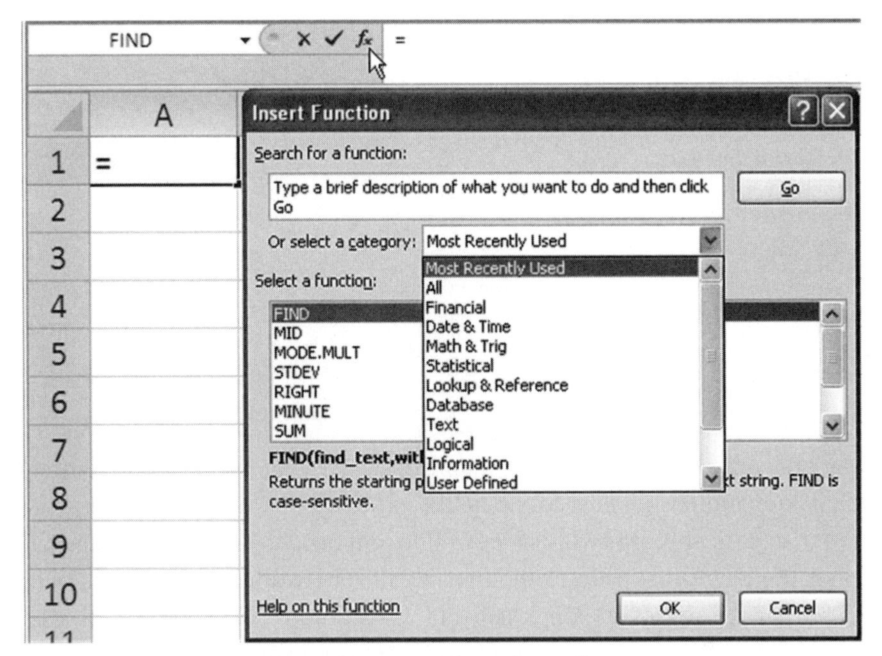

Figure 2.3 Excel formulas toolbar.

Let's use Excel "Text" functions to split a cell containing a first and last name into two cells. Simply click on the cell next to the full name, insert a "Text" function, and choose "MID" (Fig. 2.4).

I often find it's easiest to start with a simple formula and then enhance it to do what I want. In this case, let's select just the first name of the first person by putting A2, 1, and 3 as the parameters of the function. The "MID"

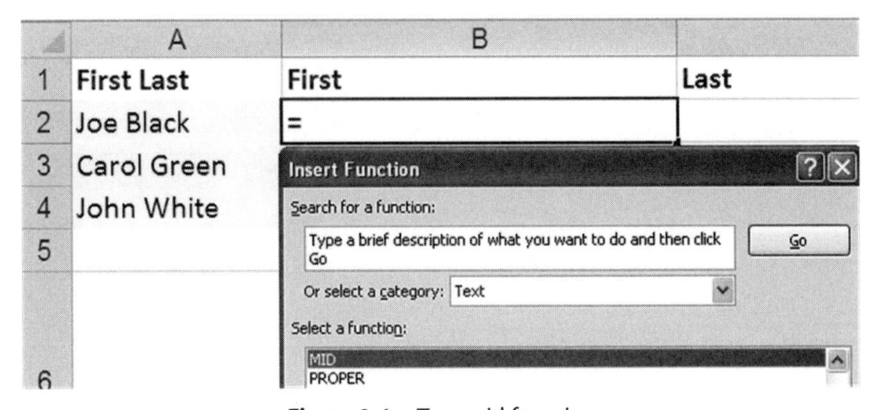

Figure 2.4 Text-mid function

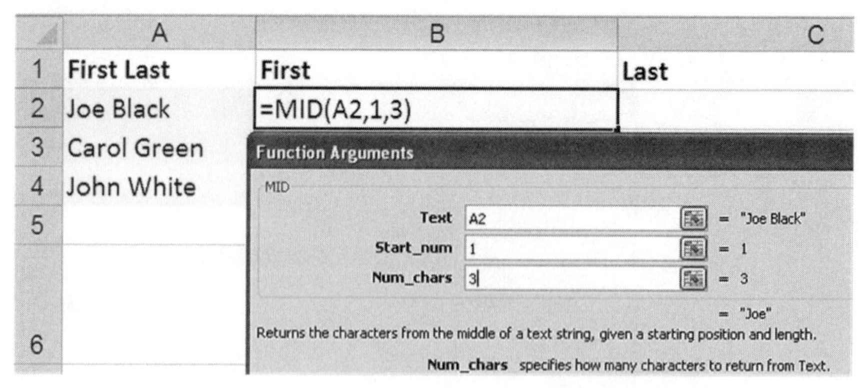

Figure 2.5 Using the text MID function with FIND.

function will select "Joe" from the name in cell A2 (Fig. 2.5). Excel will show the values and results next to the fields.

Of course, this only works for the "Joe" but not for "Carol," so we'll want to set the length to be up to the first blank. There are two functions that search for strings inside other strings ("FIND" and "SEARCH"). Let's use "FIND." Rather than trying to build this into the existing formula directly, I often use a different cell so that I can see the result (Fig. 2.6). In this case, I'm telling Excel to find a blank (" ") in A2 starting at position 1, giving the answer, 4 (the character after "Joe").

Four is one greater than the three I need, so I'll have to subtract one from the result. Now I just copy the "FIND" formula from cell D2 and insert it into the "MID" formula where the "3" used to be (Fig. 2.7). Excel will let us *nest* formulas inside other formulas to get the result.

If we copy and paste the formula for the other names, we get the first name for each person (Fig. 2.8).

	A	B	C	D
1	**First Last**	**First**	**Last**	
2	Joe Black	Joe		=FIND(" ",A2,1)
3	Carol Green			
4	John White			
5				
6				
7				

Function Arguments

FIND

Find_text " " = " "

Within_text A2 = "Joe Black"

Start_num 1 = 1

= 4

Returns the starting position of one text string within another text string. FIND is case-sensitive.

Start_num specifies the character at which to start the search. The first character in Within_text is character number 1. If omitted, Start_num = 1.

Figure 2.6 Find blank in text string.

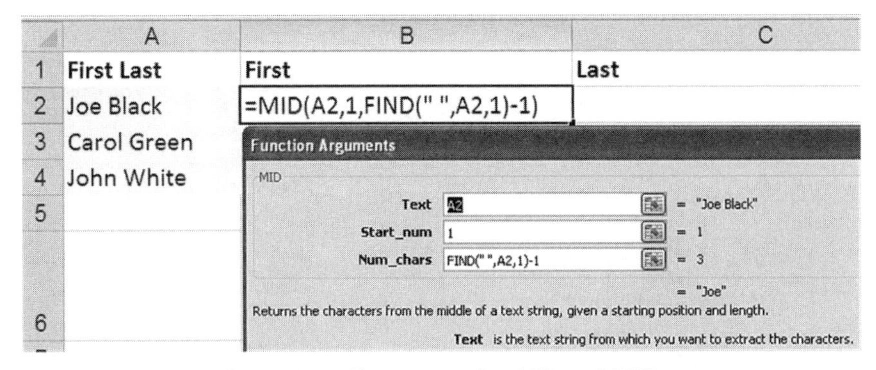

Figure 2.7 First name using MID and FIND.

B2 *fx* =MID(A2,1,FIND(" ",A2,1)-1)

	A	B
1	**First Last**	**First**
2	Joe Black	Joe
3	Carol Green	Carol
4	John White	John
5		MID(A2,1,FIND(" ",A2,1)-1)

Figure 2.8 All first names.

The formula for the last name is similar except that we use "FIND+1" as the *starting point* and use the "LEN" function to get the maximum length of the name (Fig. 2.9).

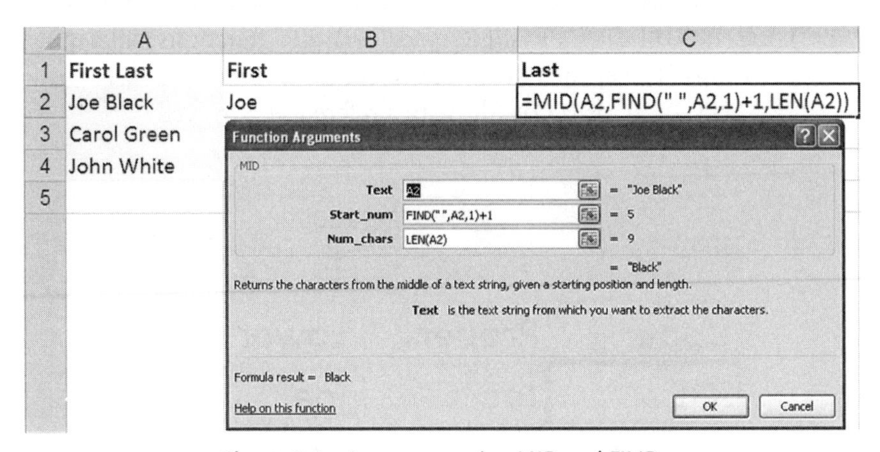

Figure 2.9 Last name using MID and FIND.

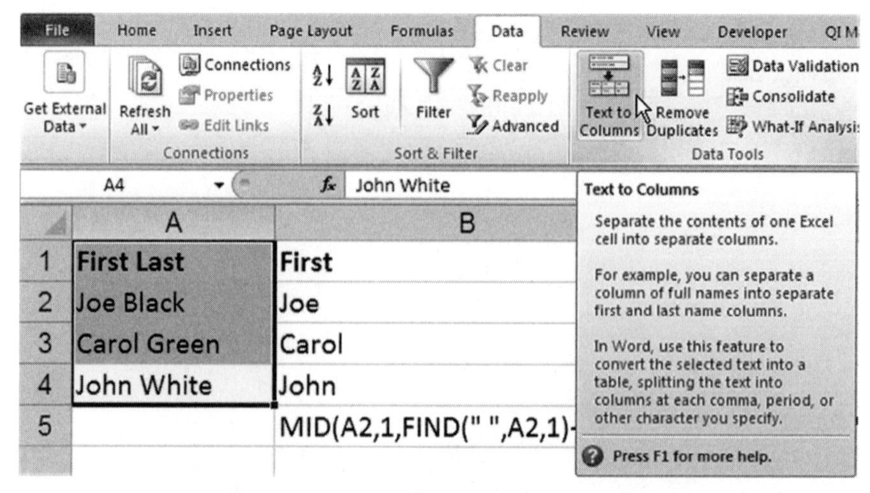

Figure 2.10 Excel's Text to Columns tool.

Now we've effectively split the first and last names out of the full name. While this could have been done with Excel's "Text to Columns" function (Fig. 2.10), it won't always work with the crazy ways data gets into cells. Sometimes the only way to dig out the data you need is with "MID" and "FIND" or "SEARCH."

Upper–Proper–Lower

Some computer systems store text in uppercase; some in lowercase. Some people enter names in lowercase ("joe") when they should be capitalized ("Joe"). Excel's "UPPER," "PROPER," and "LOWER" functions will handle these situations. To get Joe's name in the correct format (Fig. 2.11), use "UPPER," "PROPER," or "LOWER" with a reference to the cell.

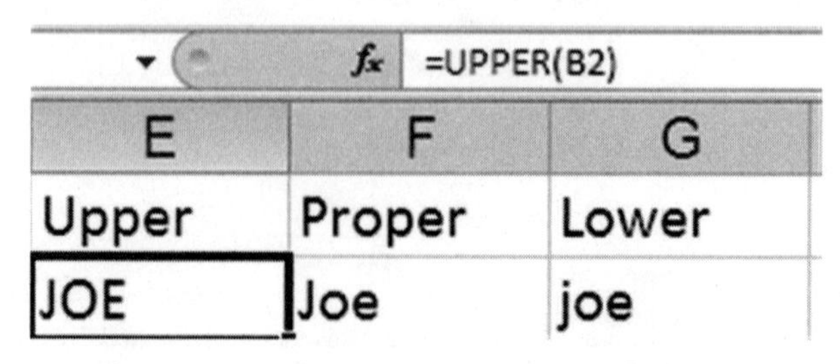

Figure 2.11 UPPER-PROPER-LOWER functions.

Time and Date Formulas

Time or date data is another format that people can tweak to their needs. If you want to know the difference between two times (e.g., a start and end time), simply input a formula, "=B2-A2," to subtract one from the other (Fig. 2.12) to get the elapsed time.

C2 fx =B2-A2

	A	B	C
1	Start Time	End Time	Subtract Start from End Time
2	5/18/13 9:33	5/18/13 9:53	0:20

Figure 2.12 Calculating elapsed time.

Then, if you want to convert the time or date format into minutes, you will need a formula that converts days, minutes, and seconds into a single value. There are 1,440 minutes in a day and 60 in an hour. "DAYS360," "HOUR," and "MINUTE" convert dates or times into a count you can multiply and add to get the result you want (Fig. 2.13). (*Note:* I used "DAYS360" because the "DAY" function maxes out at 31 days and then starts over at 1, which could be a problem.)

What if we want to convert times or dates into hours? The formula is very similar (Fig. 2.14). This gives us the elapsed time in various formats that can be easily graphed using the QI Macros.

D2 fx =DAYS360(A2,B2)*1440+HOUR(C2)*60+MINUTE(C2)

	A	B	C	D
1	Start Time	End Time	Subtract Start from End Time	Convert Time in C2 to Minutes
2	5/18/13 9:33	5/18/13 9:53	0:20	20

Figure 2.13 Elapsed time in minutes.

E2 fx =DAYS360(A2,B2)*24+HOUR(C2)+MINUTE(C2)/60

	A	B	C	D	E
1	Start Time	End Time	Subtract Start from End Time	Convert Time in C2 to Minutes	Convert Time in D2 to Hours
2	5/18/13 9:33	5/18/13 9:53	0:20	20	0.33

Figure 2.14 Elapsed time in hours.

If you just want the month, day, or year from a date, use "MONTH," "DAY," or "YEAR" functions (Fig. 2.15).

fx =MONTH(A2)

H	I	J
Start Month (A2)	Start Day (A2)	Start Year (A2)
5	18	2013

Figure 2.15 MONTH DAY YEAR functions.

If you just want to combine a month, day, and year into a date, use the "DATE" function (Fig. 2.16).

If you just want the current date, use the "TODAY" function (Fig. 2.17).

L2 fx =DATE(J2,H2,I2)

	H	I	J	K	L
1	Start Month (A2)	Start Day (A2)	Start Year (A2)		Convert M/D/Y to Date
2	5	18	2013		5/18/2013

Figure 2.16 DATE function.

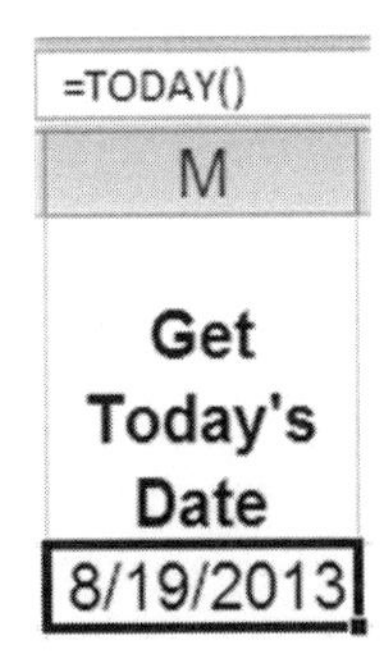

Figure 2.17 Today's date.

If you just want the current date and time, use the "NOW" function (Fig. 2.18).

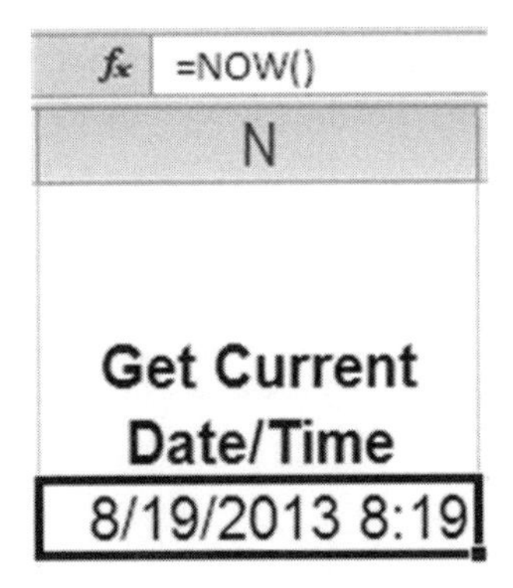

Figure 2.18 Current date and time.

Converting Times to Hours

One hospital operating room was trying to determine how long it takes from the time a patient arrives until he or she enters the operating room. The hospital had the actual times, so the staff could have run a histogram, but they wanted to categorize it as 0–1 hour, 1–2 hours, and so on.

The formula to convert time to an hour span involves concatenating the hour with a dash (–) and some addition (Fig. 2.19). So 2 hours and 44 minutes (2:44) becomes 2–3.

Then it is easy to use a PivotTable to summarize the data (Fig. 2.20) and put it in a Pareto chart (Fig. 2.21).

=HOUR(Q2)&"-"&HOUR(Q2)+1

P	Q	R	S	T
Sch Time	Arrival to Ready in OR	Hours		
7:50	2:44	=HOUR(Q2)&"-"&HOUR(Q2)+1		
7:30	2:44	2 HOUR(serial_number)		
7:30	2:44	2-3		
7:30	2:44	2-3		
8:20	3:27	3-4		
9:00	4:38	4-5		

Figure 2.19 Convert time to hour span.

	A	B
1	Count of Hours	
2	Hours	Total
3	2-3	1717
4	1-2	1259
5	3-4	365
6	4-5	106
7	5-6	33
8	0-1	32
9	6-7	6
10	7-8	4
11	8-9	1
12	9-10	1
13	Grand Total	3524

Figure 2.20 PivotTable of times.

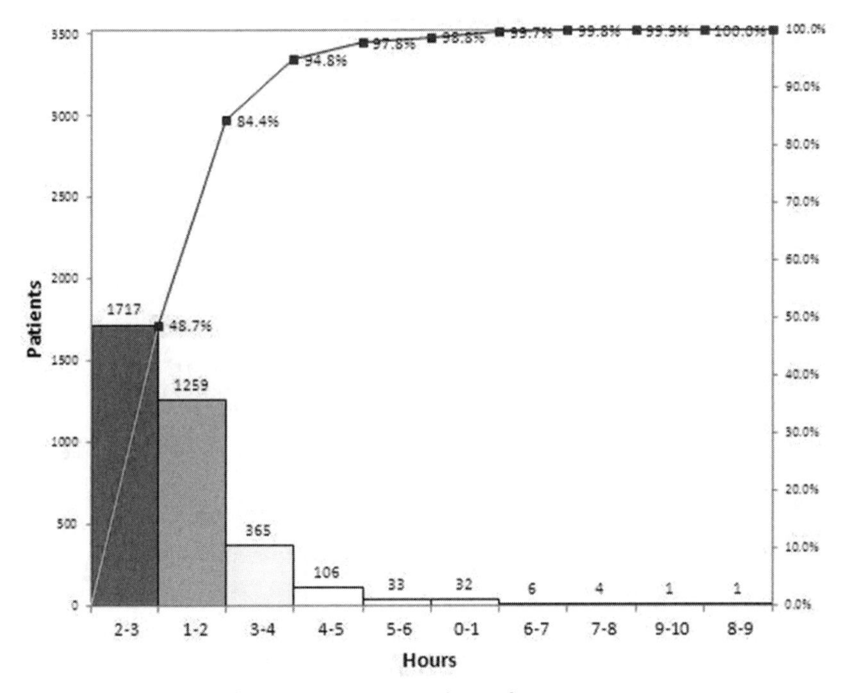

Figure 2.21 Pareto chart of time to OR.

Logic

Sometimes we need IF-THEN-ELSE logic. Recently, a client was trying to figure out how to evaluate a report to show pass/fail so that it could be counted with a PivotTable. To do this, we had to add some formulas. In this case, a certain value "X" had to be between two values. The first formula evaluates whether Prod1's "X" value is less than 0.05 and puts the word "Pass" or the word "Fail" in the cell (Fig. 2.22).

FIND | =IF(B3<0.05, "Pass","Fail")

	A	B	C	D	E
1	Product	X	Pass/Fail	Criteria	
2	Prod1	0.03	Pass	<0.05	
3	Prod1	0.01	=IF(B3<0.05, "Pass","Fail")		

Figure 2.22 Analyzing pass-fail criteria.

	A	B	C	D
1	Product	X	Pass/Fail	Criteria
2	Prod1	0.03	Pass	x<0.05
3	Prod1	0.01	Pass	x<0.05
4	Prod1	0.10	Fail	x<0.05
5	Prod2	1.00	=IF(A5="Prod1",IF(B5<0.05,"Pass","Fail"),	.90<x<1.1
6	Prod2	0.94	IF(A5="Prod2",IF(AND(0.9<B5,B5<1.1),	.90<x<1.1
7	Prod3	3.00	"Pass","Fail"),IF(A5="Prod3",IF(AND(1.9<	1.9<x<2.1
8	Prod3	2.00	B5,B5<2.1),"Pass","Fail"))))	1.9<x<2.1
9	Prod3	1.97	P IF(logical_test, [value_if_true], [value_if_false])	1.9<x<2.1

Figure 2.23 Pass-fail for multiple products.

Of course, this formula won't work for all the products in the report that have other specification limits. Thus we had to expand the formula to check the product name and choose the right limits (Fig. 2.23).

The pass/fail results now can be counted easily with a PivotTable.

If we have too many products, we might want to use a lookup formula with some specification limits (Fig. 2.24). Lower and upper specification limits ("LSL" and "USL") are in cells I1 through K6. We can use "VLOOKUP" to find the lower and upper specification limits based on the type of product, and then compare it with the value ("X").

	F	G	H	I	J	K
1	Product	X	Pass/Fail Lookup	Product	LSL	USL
2	Prod1	0.03	Pass	Prod1	0	0.05
3	Prod1	0.01	Pass	Prod2	0.9	1.1
4	Prod1	0.10	Fail	Prod3	1.9	2.1
5	Prod2	1.00	Pass	Prod4	2.9	3.1
6	Prod2	0.94	Pass	Prod5	3.9	4.1
7	Prod3	3.00	=IF(AND(VLOOKUP(			
8	Prod3	2.00	F7,I$2:K$6,2)<G7,G7<			
9	Prod3	1.97	VLOOKUP(F7,I$2:K$6,			
10			3)),"Pass","Fail")			

Figure 2.24 Pass-fail using VLOOKUP.

Here's My Point

Sometimes data has to be manipulated to provide the right starting point for analysis. Sometimes you need a simple mathematical formula, sometimes text, sometimes IF-THEN-ELSE logic. Regardless, you will want to develop a basic grasp of Excel formulas and functions to make your life easier. Have fun exploring Excel's functions.

CHAPTER 3

Filtering Data for Breakthrough Improvement

Many Excel users find themselves faced with huge data dumps from corporate systems. And they wonder, "What do I do with this?" Sometimes the first step is to reduce the data to a manageable size using Excel's "AutoFilter" and "Sort" functions.

Create Subsets of Your Data Using "AutoFilter"

Getting data out of corporate systems into Excel is usually easy. Virtually all systems allow some sort of export to Excel. Then comes the hard part: reducing the data. A data dump might look like Figure 3.1. (If you have loaded the QI Macros or a 30-day evaluation copy, this spreadsheet is on your computer in "Documents\QI Macros Test Data\pivottable.xls.")

But maybe we only want to look at Hospital 1. You can use Excel's "AutoFilter" to reduce the data shown. To do this, click on "Data Filter" (Fig. 3.2).

	A	B	C	D	E	F	G	H	I	J	K
1	Region	POST DATE	ENT	ADM DATE	DIS DATE	AS	COS	FC	IN1	PT	DENIED CHARGES
2	North	6/27/03	Hosp1	2/13/03	1/1/00	OL		X	AEH	O	543.07
3	South	12/24/02	Hosp2	7/13/02	1/1/00	OL		X	BCP	E	215.4
4	South	2/25/03	Hosp2	12/6/02	1/1/00			X	CGH	O	157.92
5	South	5/23/03	Hosp3	10/20/02	1/1/00	OL		X	MAH	O	90.73
6	North	7/15/03	Hosp1	5/7/03	1/1/00	AP		X	HEH	O	4103.78
7	North	11/5/02	Hosp4	8/6/01	1/1/00	OL		F	PTB	E	3224.83
8	North	11/20/02	Hosp5	4/15/02	1/1/00	OL		F	PTB	O	3291.76
9	North	11/27/02	Hosp1	5/13/02	1/1/00	OL		F	PTB	O	13845.9
10	North	11/27/02	Hosp4	9/16/02	1/1/00			F	PTB	O	1151

Figure 3.1 Corporate data dump.

	B	C	D				H	I	J	K
1	POST DATE	ENT	ADM DATE					IN1	P	DENIED CHARGES
2	6/27/03	Hosp1	2/13/03					AEH	O	543.07
3	12/24/02	Hosp2	7/13/02					BCP	E	215.4
4	2/25/03	Hosp2	12/6/02	1/1/00			X	CGH	O	157.92
5	5/23/03	Hosp3	10/20/02	1/1/00	OL		X	MAH	O	90.73
6	7/15/03	Hosp1	5/7/03	1/1/00	AP		X	HEH	O	4103.78
7	11/5/02	Hosp4	8/6/01	1/1/00	OL		F	PTB	E	3224.83
8	11/20/02	Hosp5	4/15/02	1/1/00	OL		F	PTB	O	3291.76
9	11/27/02	Hosp1	5/13/02	1/1/00	OL		F	PTB	O	13845.9
10	11/27/02	Hosp4	9/16/02	1/1/00			F	PTB	O	1151

Figure 3.2 Autofilter the data.

Then click on the pull-down arrow next to "ENT," and select "Hosp1" (Fig. 3.3).

This will select only the data that contains "Hosp1" (Fig. 3.4).

Notice that Excel changes the icon of the pull-down button to show that the column has been filtered (Fig. 3.5).

Figure 3.3 Select hospital 1.

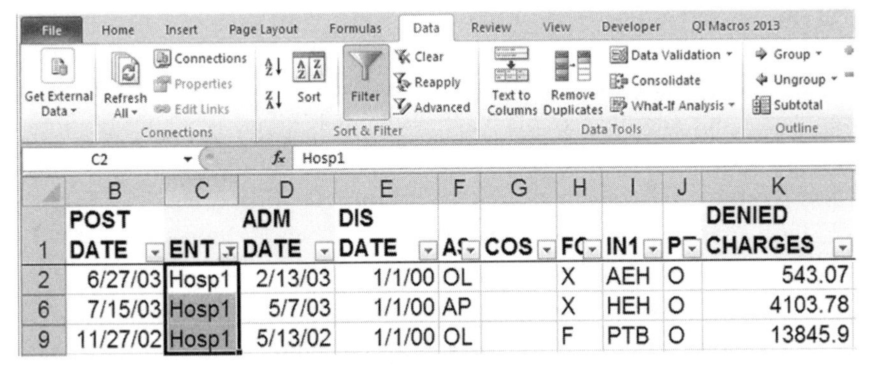

Figure 3.4 Hospital 1 data.

Figure 3.5 AutoFilter icon.

To remove the selection, simply click on the pull-down menu and "Select All" (Fig. 3.6).

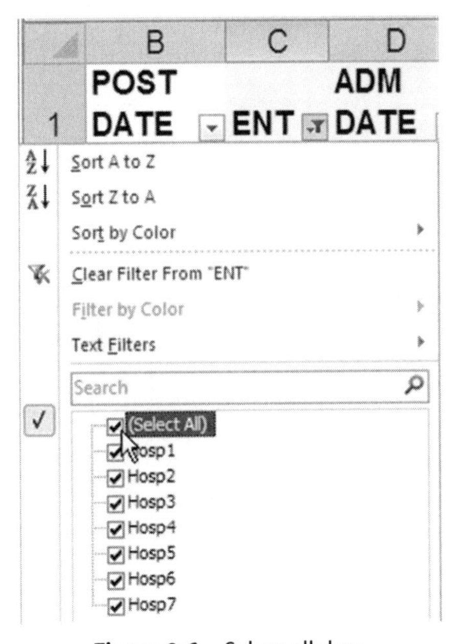

Figure 3.6 Select all data.

Custom Filters

Alternatively, you might want to do a custom selection. Let's say that we only want to see the patient accounts that have denied charges of $2,000 or more. To do this, click on the pull-down arrow next to "DENIED CHARGES," and select "Number Filters–Greater Than Or Equal To" (Fig. 3.7).

This opens a dialog window (Fig. 3.8). Change the selection to "is greater than or equal to," and set the value to 2000.

Then click "OK" to see the data on the balances over $2,000 (Fig. 3.9).

If you only want to see the balances for Hospital 1 that are over $2,000, click on "ENT," and select "Hosp1" (Fig. 3.10).

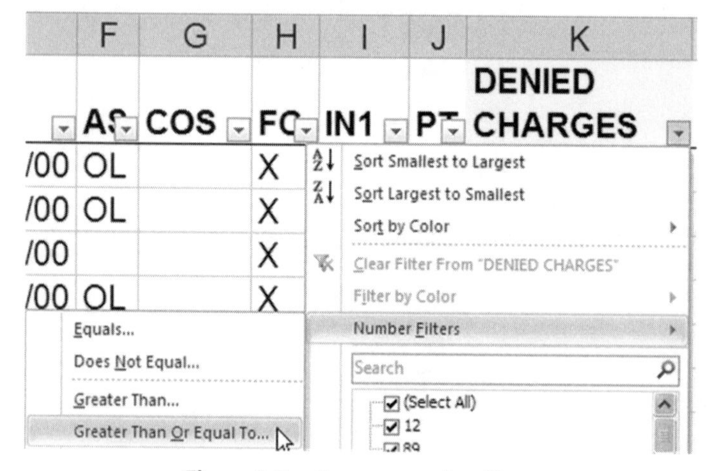

Figure 3.7 Custom number filter.

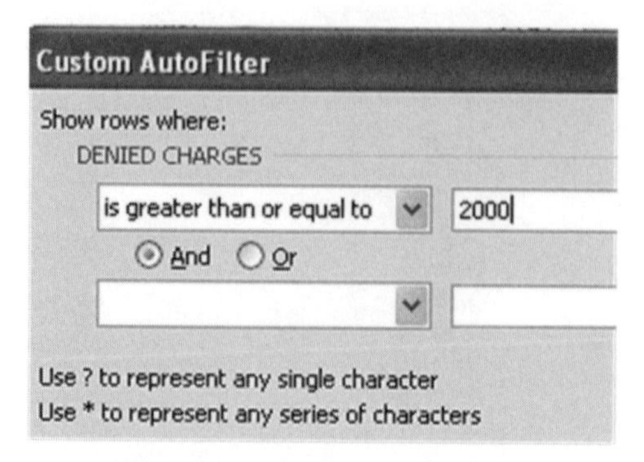

Figure 3.8 Custom AutoFilter menu.

	B	C	D	E	F	G	H	I	J	K
1	POST DATE	ENT	ADM DATE	DIS DATE	AC	COS	FC	IN1	PT	DENIED CHARGES
6	7/15/03	Hosp1	5/7/03	1/1/00	AP		X	HEH	O	4103.78
7	11/5/02	Hosp4	8/6/01	1/1/00	OL		F	PTB	E	3224.83
8	11/20/02	Hosp5	4/15/02	1/1/00	OL		F	PTB	O	3291.76
9	11/27/02	Hosp1	5/13/02	1/1/00	OL		F	PTB	O	13845.9
13	6/25/03	Hosp5	7/26/01	1/1/00	OL		F	PTB	O	2124.86
18	9/11/03	Hosp6	2/4/03	1/1/00	OL		F	PTB	O	3728
26	1/2/03	Hosp2	3/4/02	1/1/00	OL		X	BAN	O	2299.71
32	1/10/03	Hosp3	3/13/01	1/1/00	OL		X	HEH	R	6908.98
33	3/27/03	Hosp2	1/14/03	1/1/00			X	OHP	R	4512.11
36	2/13/03	Hosp2	5/6/02	1/1/00	OL		X	RMH	S	11045.57

Figure 3.9 Denied charges over $2,000.

	B	C	D	E	F	G	H	I	J	K
1	POST DATE	ENT	ADM DATE	DIS DATE	AC	COS	FC	IN1	PT	DENIED CHARGES
6	7/15/03	Hosp1	5/7/03	1/1/00	AP		X	HEH	O	4103.78
9	11/27/02	Hosp1	5/13/02	1/1/00	OL		F	PTB	O	13845.9

Figure 3.10 Denied charges for hospital 1 over $2,000.

As you can see, you can continue to select and subset the data in any way you choose. You can copy and paste these subsets into another spreadsheet *without* any of the hidden rows using "Edit–Copy/Paste." You cannot, however, paste data back into the autofiltered sheet without overwriting hidden rows.

If you get lost and want to start over, simply click on "Data Filter–Select All" (see Fig 3.6).

Sort the Data

Excel's "Sort" function can help to organize the data for clarity. Date and text fields can be sorted using Excel's "Data" tab and the *ascending-order button* (Fig. 3.11). Just click in the column you want sorted, and click the ascending-order button. Excel will automatically expand the selection to include all the data.

Because we want to focus on the *worst problems first*, sort numeric values such as "DENIED CHARGES" in *descending order* (Fig. 3.12).

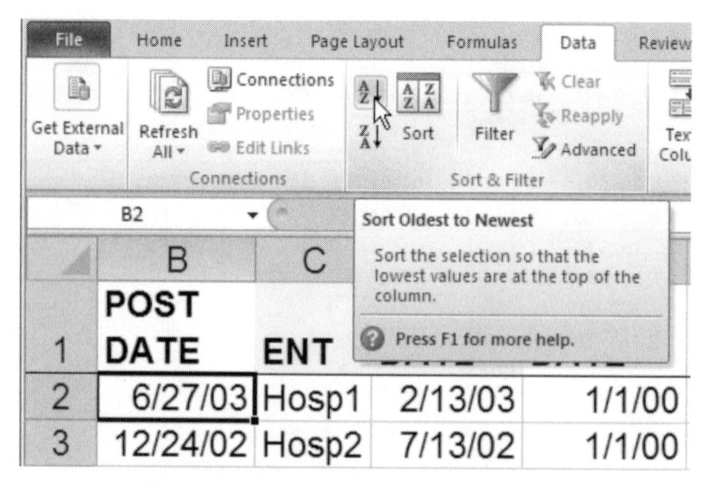

Figure 3.11 Sort data in ascending order.

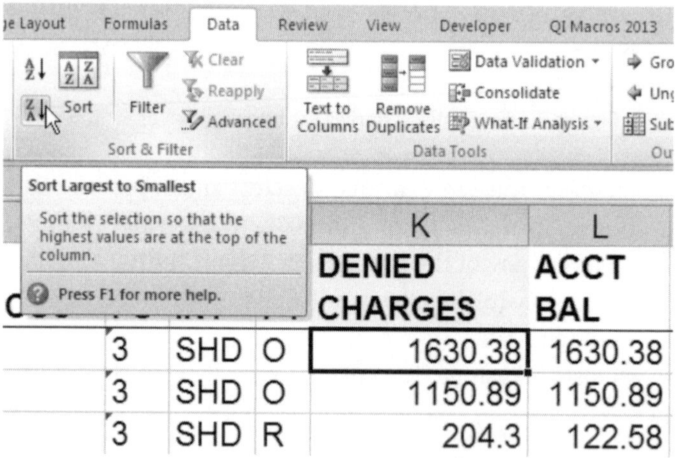

Figure 3.12 Sort data in descending order.

Which problems would be more valuable to solve—the top five denied claims worth \$40,000 or the bottom five denied claims worth less than \$500 (Fig. 3.13). To achieve breakthrough improvements, it's essential to focus on the highest leverage points with the greatest payoffs. *First the whales; then the minnows. Fix the worst first!*

	C	D	E	F	G	H	I	J	K
1	ENT	ADM DATE	DIS DATE	AS	COS	FC	IN1	PT	DENIED CHARGES
2	Hosp1	5/13/02	1/1/00	OL		F	PTB	O	13845.90
3	Hosp2	5/6/02	1/1/00	OL		X	RMH	S	11045.57
4	Hosp3	3/13/01	1/1/00	OL		X	HEH	R	6908.98
5	Hosp3	7/14/02	1/1/00	OL		7	SHF	E	5967.74
6	Hosp2	11/15/02	1/1/00	OL		X	RMH	S	4535.66
47									
48	Hosp3	6/19/02	1/1/00			7	SHF	O	100.19
49	Hosp3	10/20/02	1/1/00	OL		X	MAH	O	90.73
50	Hosp6	8/8/03	1/1/00			F	PTB	O	89
51	Hosp6	8/8/03	1/1/00			F	PTB	O	89
52	Hosp4	6/2/03	1/1/00			F	PTB	O	12

Figure 3.13 Largest versus smallest denied charges.

Even Huge Spreadsheets Can Be Manageable

Excel's "AutoFilter" can help to reduce the amount of data served up by corporate systems to a manageable level. Excel's "Sort" function can help to identify the greatest pain points and greatest opportunities. From there, you can:

- Use Excel's "PivotTable" or QI Macros "PivotTable Wizard" to summarize the selected data.
- Use the QI Macros to graph the summarized data.

Use the power of Excel to narrow your focus to the few key things you want to understand, diagnose, or improve. Then use the QI Macros to graph the results and present your business case (i.e., improvement story) to the decision makers.

CHAPTER 4

The Magnificent Seven Tools for Breakthrough Improvement

The 1960s film *The Magnificent Seven* starred Yul Brenner, Steve McQueen, Charles Bronson, James Coburn, Robert Vaughn, Brad Dexter, and Horst Buckholz. These seven hired gunmen protect a Mexican village from the bandit Calvera played by Eli Wallach. The film was an Americanization of the Japanese film *The Seven Samurai*. The film implies that you don't need an army to win a war, just seven top gunmen.

The Magnificent Seven Tools

> Becoming a master of karate was not about learning 4,000 moves but about doing just a handful of moves 4,000 times.
>
> —CHET HOLMES

Over the years, in project after project, I have found myself returning to the same magnificent seven tools (Table 4.1).

Table 4.1 The Magnificent SevenTools

<table>
<tr><td>1.</td><td>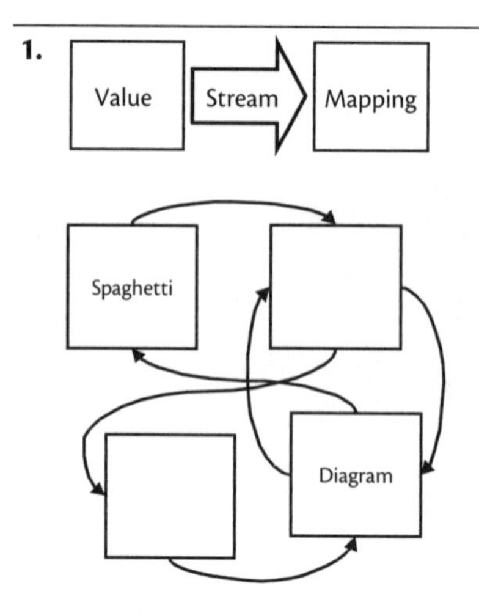
</td><td>Value-stream maps to identify and remove delays from any process (see Chap. 11).

Spaghetti diagrams to identify and eliminate unnecessary movement of people or materials in any workspace.</td></tr>
<tr><td>2.</td><td>
</td><td>Excel PivotTables to summarize massive dumps of data about defects and money. (Every multimillion-dollar improvement project I've ever worked on began with PivotTables; see Chap. 9.)</td></tr>
<tr><td>3.</td><td>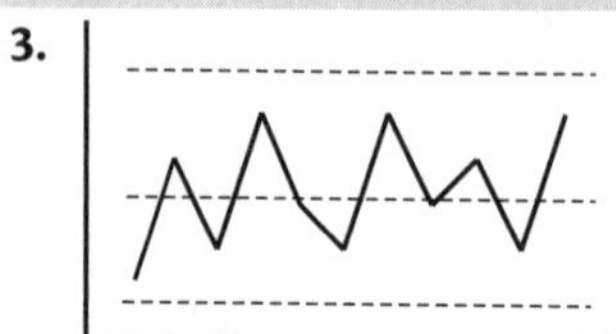</td><td>Control charts to measure and monitor the performance of any process (see Chap. 10). (Without control charts, it is impossible to detect changes in an improved process until they become epidemic or catastrophic.)</td></tr>
<tr><td>4.</td><td>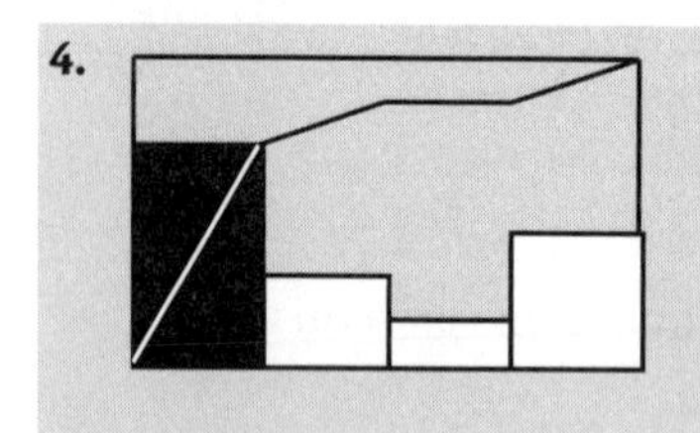</td><td>Pareto charts to identify the most frequent types of defects, mistakes, and errors. (Breakthrough improvement requires that we fix the worst first. Pareto charts can quickly identify where to focus improvement efforts for maximum benefit.)</td></tr>
<tr><td>5.</td><td>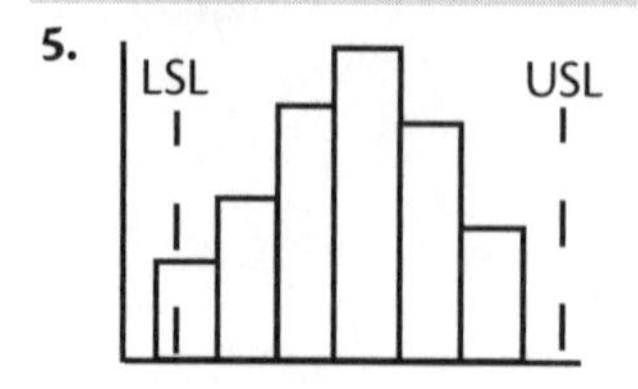
</td><td>Histograms to evaluate deviations in process performance (see Chap. 10). (Is the part or product too big or too small, too long or too short, or too wide or too thin? Is the time to repair too long? Histograms can answer these questions.)</td></tr>
</table>

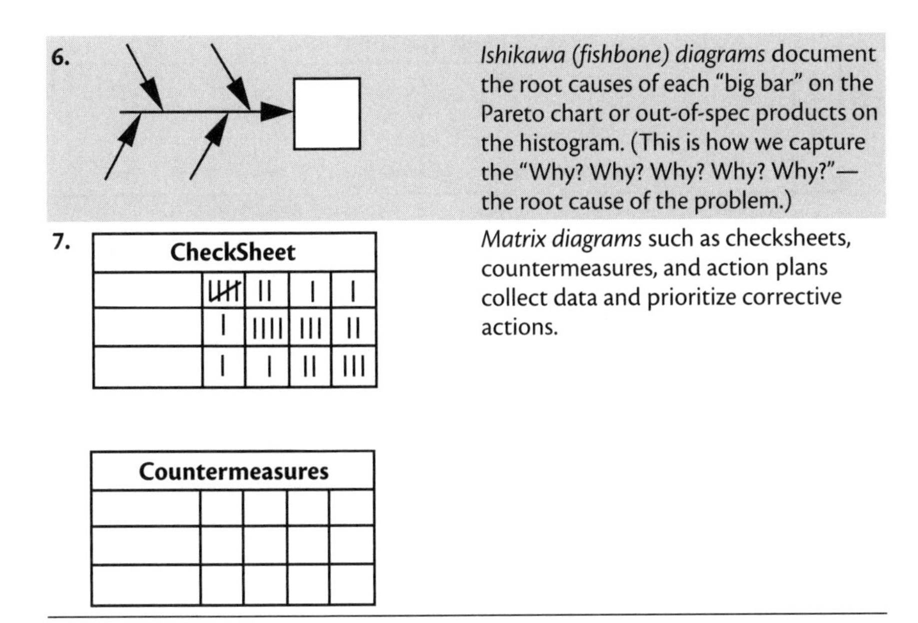

I have found that these seven tools can help teams solve 99 percent of the problems facing a typical business. Sure, you will need more exotic tools to solve problems in the last 1 percent, but you don't need them for a while. Quite a while.

Not sure what chart to choose? Just select your data, and click on the QI Macros "Chart Wizard." It will examine your data and draw every chart it can think of using the data provided.

Are you stuck trying to figure out what tools to use in what order? Learn the magnificent seven to deliver breakthrough improvements (Fig. 4.1). You'll be surprised at how far you can go with just these tools.

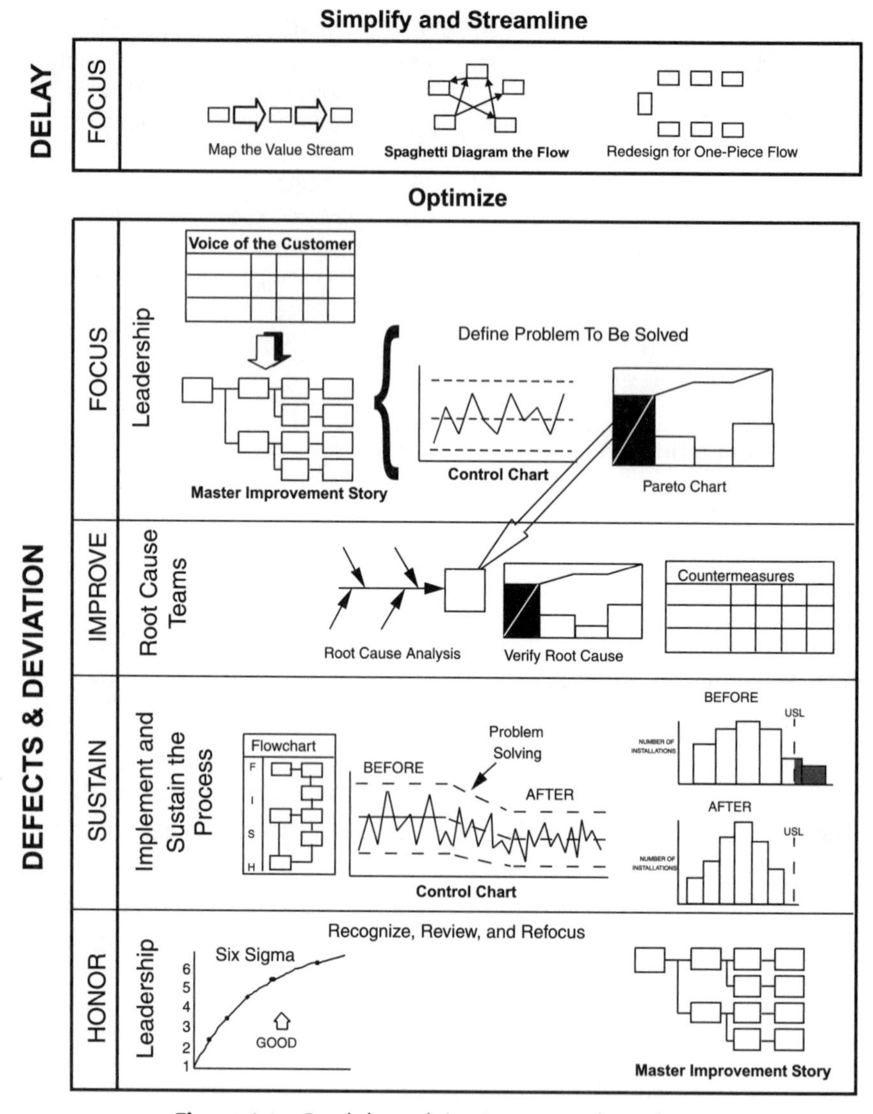

Figure 4.1 Breakthrough improvement cheat sheet.

CHAPTER 5

Follow the Money for Breakthrough Improvement

I think we can all agree that both service and manufacturing companies make too many mistakes and errors, have too many unnecessary delays, make products that aren't quite as easy to assemble as we'd like, and generally irritate customers at every turn. This means that suppliers give your company things that aren't quite right that you then turn into products and services that aren't quite right. The sales department gives the production department orders that are incomplete and inaccurate. Production delivers information to billing that is similarly incomplete and inaccurate and products or services that make the customer call for support. It's a giant cycle of mistakes and errors, waste and rework that consume *a third or more of every company's revenues.* And yet companies still make a profit! So how do we start to solve problems in businesses large and small, service and manufacturing using QI Macros and Excel? When in doubt, act like a homicide detective: *follow the money.*

Meet Sarah. Sarah has a booming landscaping business. Homeowners and corporate and government building owners all use her services. Her employees are doing well, but Sarah is barely getting by (Fig. 5.1). Income barely exceeds expenses. She knows that she either needs to make more money or reduce expenses. But business is already booming, and she needs every employee to meet the demand. How is she going to be able to do either?

Fortunately, Sarah has a friend who is an expert in breakthrough improvement using the QI Macros and Excel. Because Sarah is already overwhelmed with work, Adrienne focuses on expenses. Rather than focus on employees, she focuses on the plants, shrubs, trees, sod, and rock that Sarah uses. It doesn't take her long to figure out that Sarah *simply isn't billing customers* for everything she plants.

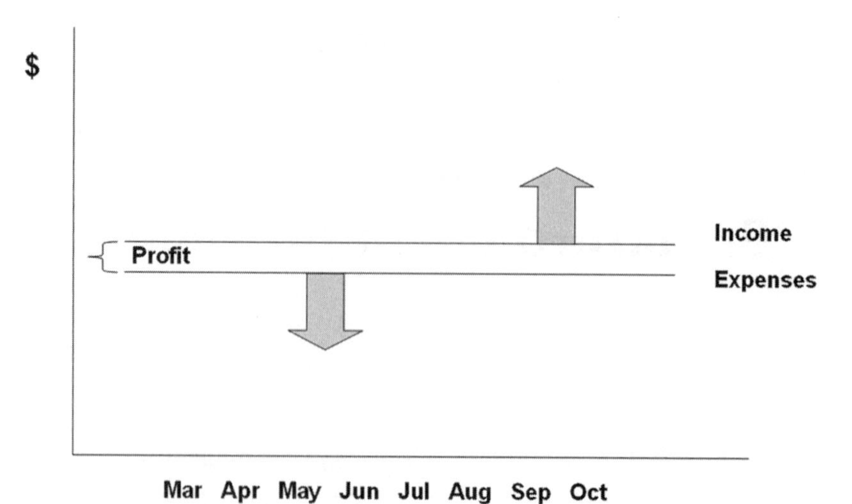

Figure 5.1 Increase profits or reduce expenses.

Sheepishly, Sarah admits that she usually puts in a few more plants here and there to make the landscape look beautiful, but she doesn't bill for them because they aren't in the estimate. Sarah's customers love her work and are more than happy to pay, but Sarah is underestimating. Adrienne helps Sarah to tweak her estimating process to cover her expenses, and she helps Sarah to develop a worksheet to track how many plants are used and reconcile them with the number purchased.

Sarah's profits soar! She is thrilled, but Adrienne isn't done yet.

Another significant expense involves replacing plants that die. Not only do they need to be repurchased, but there is also labor involved in replacing them—a classic case of unnecessary rework. A simple Pareto analysis finds that most of these dead plants come from a single supplier. When Sarah shifts her purchases to her other two suppliers, not only do her rework costs go down, but she also gets a better deal because of increased purchase volume.

Sarah's profits soar some more.

The Lesson Is Simple: Follow the Money!

Printers may have to recycle too much printed paper. Metal manufacturers may put too much out-of-spec product back into the furnace. Hospitals may have to take care of patients who fall and break a hip while in their care. Restaurants may have to remake too many meals that were incorrect. Maybe a company has too many billing adjustments or refunds.

Where is your business leaking cash? Follow the money. Then plug the leaks in your cash flow.

Health Care Case Study

While much of health care quality is focused on clinical care, there's a lot that can be done with the financial side of the business to eliminate costs and increase profits.

One of my health care clients had denied insurance claims worth millions. A hospital can have great clinical success but terrible financial issues. Both the clinical and operational sides of the hospital have to work flawlessly to reduce costs and maximize patient satisfaction and outcomes.

I began by grouping insurance-claim problems into the categories of rejects, appeals, and denied, but to make these categories actionable, I had to dive a little deeper.

Because denied claims involved "real" dollars, I did a number of Pareto charts to look for more important categories. The biggest category of denied claims was for timely filing (within 45 days). Then I categorized the denied claims by insurer. When I did, there was a big surprise: one small insurer accounted for 64 percent of denied claims.

My team developed a manual work-around for this one insurer on Friday afternoon and implemented it on Monday morning. The hospital saved $5 million over the next year.

Companies are doing almost everything well. It's just a small part of the business that's causing most of the problem. This brings us to what I call the *4-50 Rule*.

The 4-50 Rule

When I was growing up in Tucson, Arizona, my family had what was then a brand-new 1955 Chevy. It was two-toned: black and white. In Tucson, the temperatures climb to over 100°F starting in May and stay that way until September. We had what my father lovingly called "2-50 air conditioning"—two windows down and 50 miles per hour. Later in life, as I worked with more and more improvement teams, the numbers I saw in the data began to suggest something similar.

Everyone's familiar with Pareto's 80/20 rule: 20 percent of your business creates 80 percent of the problem. However, most people mistakenly believe that problems are evenly distributed in that 20 percent. Why? If you're willing to believe the 80/20 rule, why would you believe that problems are evenly distributed across the 20 percent?

What I've found in working with hundreds of teams is that the same distribution applies within the 20 percent: 4 percent of your business creates over 50 percent of the problems (the 4-50 Rule). This 4 percent is the hidden gold mine in your business. Fixing the 4 percent is the secret to breakthrough improvement. Don't waste your time on anything else. If you want to plug the leaks in your cash flow to reach new heights of performance and profitability, use your data to find and fix these small but costly issues.

Fix the Biggest Category First

In the health care system, we worked through problems, category by category, using what I call the *dirty-30 process* to identify and fix each of the most common problems first. These fixes often reduced other related categories as well. While most teams want to boil the ocean or solve world hunger, when you restrict yourself to fixing the biggest category first, you'll find it easier to make a difference, and a surprising amount of benefit will come along with it.

Reducing Denied Claims in Five Days

Denied claims mean no money for services rendered because the billing process failed in some way. Nonpayment drives up the cost of health care and pushes many hospitals toward bankruptcy. In this case study, denials were over $1 million a month (Fig. 5.2).

Using Pareto Charts of Denials

Using Excel PivotTables and the QI Macros, it was easy to narrow the focus to a few key areas for improvement: timely filing (Fig. 5.3) and one insurer (Fig. 5.4).

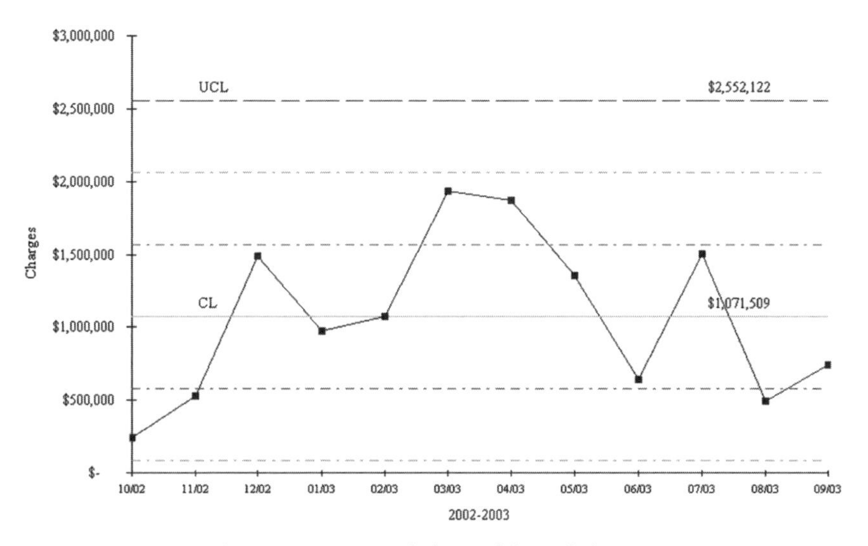

Figure 5.2 Control chart of denied charges.

Analyze Root Causes and Initiate Countermeasures

In a half-day root-cause analysis session, the team identified ways to change the process to work around the denials and change the contract process to (1) reduce delays that contribute to timely filing denials and (2) work with the insurer to resolve excessive denials.

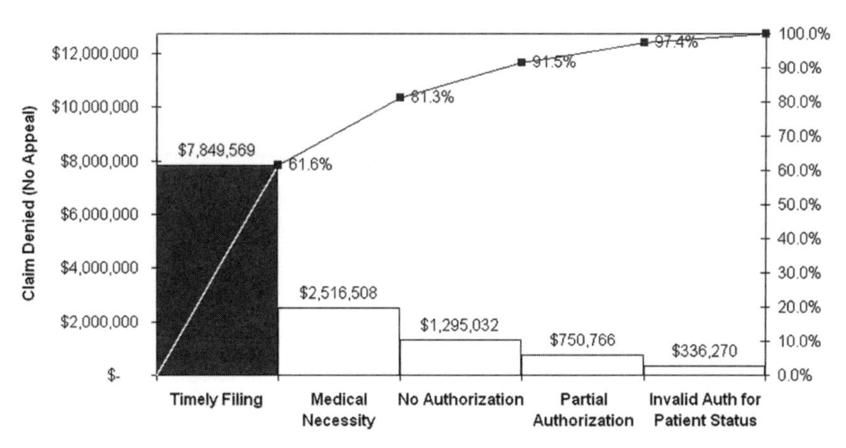

Figure 5.3 Denied claims—timely filing Pareto.

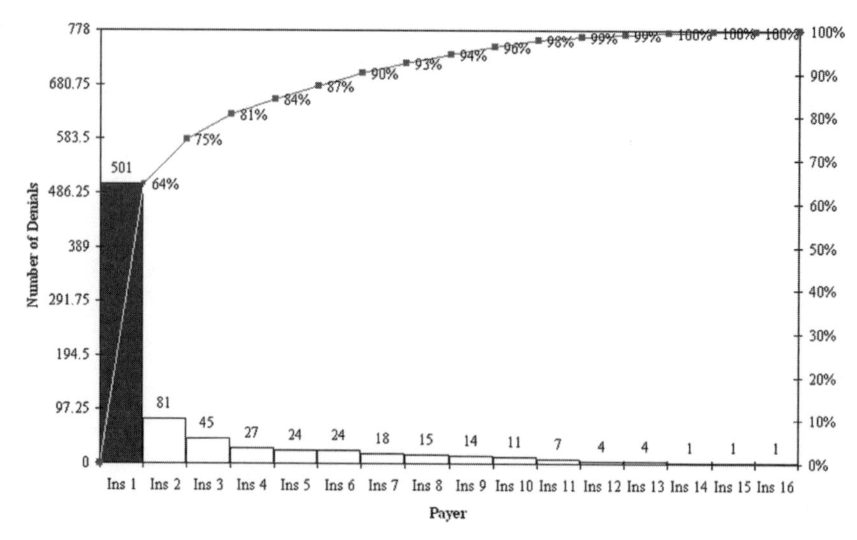

Figure 5.4 Timely filing by payer Pareto chart.

Verify Results

After implementing the process changes, denied claims fell by $380,000 per month ($5 million per year). An XmR chart (Fig. 5.5) shows denials before and after improvement.

If you go to www.qimacros.com/breakthrough, you can download the QI Macros for Excel 90-day trial. Use the "PivotTable Wizard" to create

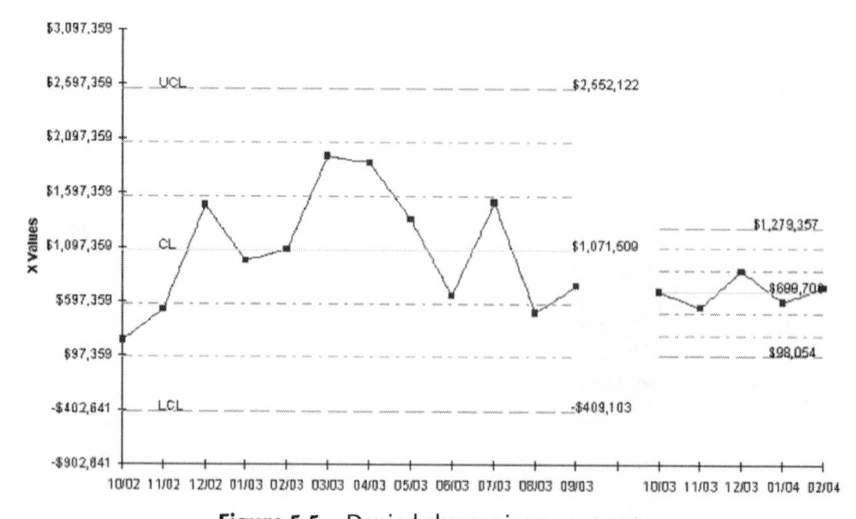

Figure 5.5 Denied charges improvement.

PivotTables. Use the "Control Chart Wizard" to draw control charts. Use the QI Macros to draw Pareto charts.

Don't Wait For Storm Season

Every business is plagued by occasional storms that are fanned by a series of problems that happen frequently enough to penetrate the dense fog of consciousness, but why wait? You already have the data you need to start finding and fixing the major categories of misses, mistakes, and errors in all aspects of your facility.

Don't just focus on the operational side (e.g., manufacturing or service), turn your attention to the big problems in the transactional side of the house: purchasing, billing, and so on. A recent study estimated that *8 of every 10 hospital bills have mistakes*. There's a whole class of consultants who, for a fee, help patients navigate this maze to get their bills paid. The same is true of restaurants, supermarkets, and retail stores. Have you ever gone to a supermarket, picked up an item on sale only to find that the scanner at checkout doesn't know anything about the discounted price?

It's not enough to provide high-quality products or services; you also must provide a high-quality experience in all aspects of service delivery, from marketing to sales to fulfillment and beyond. Customers no longer tolerate inaccurate invoices because they have to staff a group to fix them. Your customer has become part of your "fix-it factory."

Millions of people spend their lives fixing these types of errors. And they all think they're doing meaningful work, not just fixing things that shouldn't be wrong. Eliminate your fix-it factory to maximize productivity and profits.

Conclusion

Until you get to where you can prevent all the delays, defects, and deviation that siphon off your profits, become a homicide detective. *Follow the money!*

CHAPTER 6

Where's the Pain?

I am no longer surprised by the number of people who say, "I'm not sure how to apply breakthrough improvement to health care, telecom, hotels, food, transportation, [*insert your service industry here*]." It's not that difficult, but that's what I do most of the time: help people find the "pain" in their business and translate that pain into a business case for action.

While manufacturing focuses on reducing variation, services can best benefit from focusing on eliminating defects, mistakes, and errors in the service-delivery process. Because over 90 percent of people work in the service part of any business (including manufacturing), this seems like the best place to start.

Where's the Pain?

The first step is to identify the pain caused by defects, mistakes, and errors. Do these problems come from:

- Incorrect or inaccurate orders that lead to incorrect service delivery that leads to waste, rework, and lost profits?
- Inaccurate bills or invoices or improperly applied payments?
- Defects in delivered software, products, or services?
- Medication errors, hospital-acquired infections, surgical compliations, ventilator-associated pneumonia, repeat radiology or lab tests, or denied insurance claims?
- Preventable adjustments and credits?

Where's your pain?

The Dirty-30 Process for Breakthrough Improvement

While most people think that they need to start by defining and measuring something, data about the number or cost of defects, mistakes, and errors is already collected somewhere by someone. Find the data!

TAR PIT Unfortunately, the people collecting this data are often reluctant to "open their kimonos" and show it to others. Everyone has a fear of looking dumb or stupid. People fear being punished for bad results, so they are not in any hurry to share their "bad" data with anyone. I have found that once they discover that you are there to help them solve their long-standing, seemingly *unsolvable* problems, they will usually give you access to the data.

Software Case Study

I worked with one cell phone company whose pain revolved around service-order errors—there were too many of them. The company couldn't get new customers operational; it couldn't disconnect customers who switched to another provider; and it couldn't change the service to add another phone. It was a mess, and the company had 50 temporary employees correcting these service orders every day because it had a *two-month backlog* of errors. Ouch! Fortunately, this problem was so painful that the company had no problem giving me the data. The problem was in its service-order system.

I've spent over 40 years in software development and maintenance on all kinds of computers from big IBM mainframes to minicomputers to PCs and Macs. While most software improvement efforts focus on requirements, design, code, and test, because applications continue to be written by people using requirements and designs that can be flawed, software is rarely released—*it escapes.*

Information technology (IT) managers and application users often expect a new software project or enhancement release of an application to be flawless and then are stunned by the additional staffing required to stem the tide of inaccurate or rejected transactions (i.e., defects).

The secret is to:

1. Quantify the cost of correcting these rejected transactions.

2. Understand the Pareto pattern of rejected transactions.
3. Analyze 30 rejected transactions one by one to determine the root cause.
4. Revise the requirements and modify the system to prevent the problem.

Quantify the Costs

The first step in the dirty-30 process is to identify the number of rejected transactions and the associated costs. In working with this cell phone company, my team found about a 16 percent (160,000 rejects per million orders) level of rejected service orders (Fig. 6.1).

There were over 30,000 errors per month, which, at an average cost of $12.50 to fix (temporary worker wage cost only), cost $375,000 per month. The objective was to cut this level of rejects in half by the end of the year.

Understand the Pareto Pattern

All systems have routines to accept, modify, or reject incoming transaction data. These are assigned error codes and dumped into error buckets to await correction. The company actually built another application to handle much

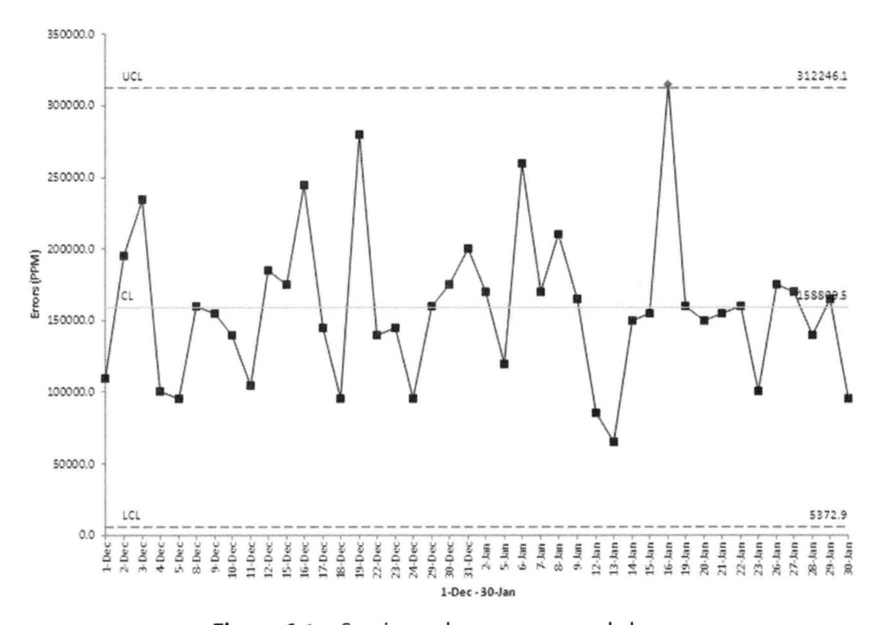

Figure 6.1 Service order errors control chart.

of the correction, but it still left significant quantities of defects to be corrected manually (Fig. 6.2). Some errors affected customer records, whereas others affected the customer's service. Although smaller in quantity, service problems caused the most pain.

There were over 200 different error codes, but only six accounted for over 80 percent of the total rejected transactions. Two affected service directly (Fig. 6.3) and four affected customer records (Fig. 6.4). *Note:* Six errors out of 200 means that 3 percent of the codes accounted for over 80 percent of total errors—the 4-50 Rule in action!

It only took about three days to gather the data and isolate these error codes as the key ones to focus on.

Analyzing the Dirty 30

The next step was to convene root-cause teams to investigate 30 rejects of each error type. It took a week or more to get the right people in the room to investigate each type of error. The right people included the IT systems analyst, error-correction people, and service-order entry people. To attempt to do all six at one time with the same people would have been foolish. The errors required different subject-matter experts, and the root causes were

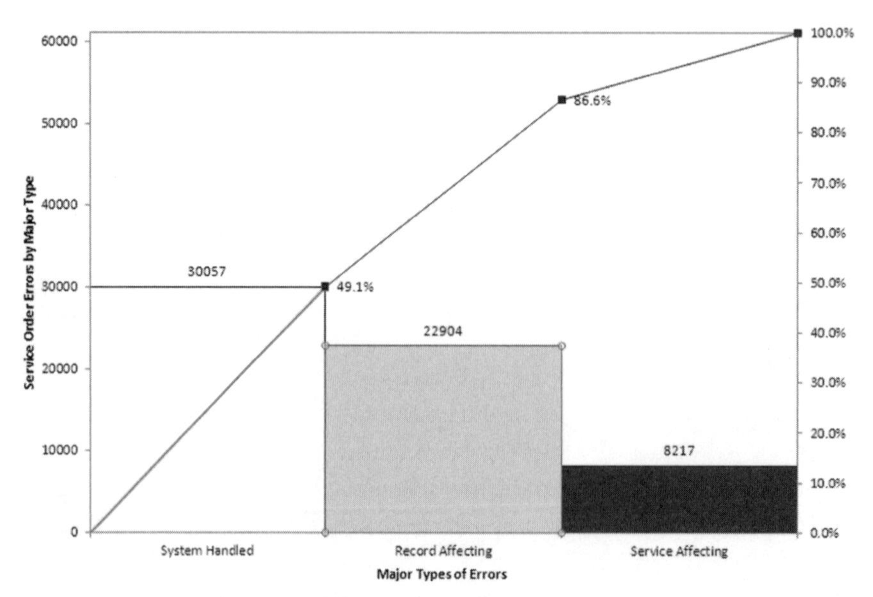

Figure 6.2 Pareto chart of major error types.

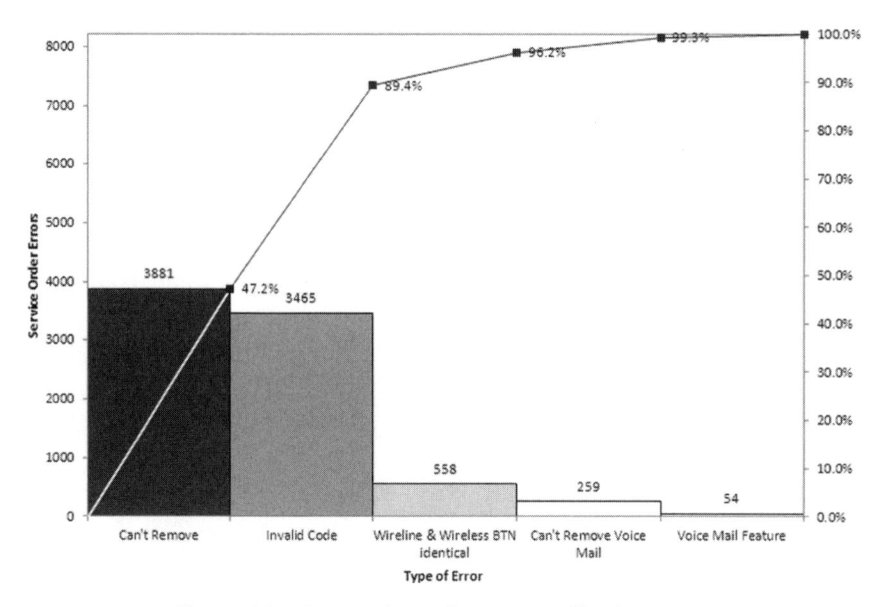

Figure 6.3 Pareto chart of customer-affecting errors.

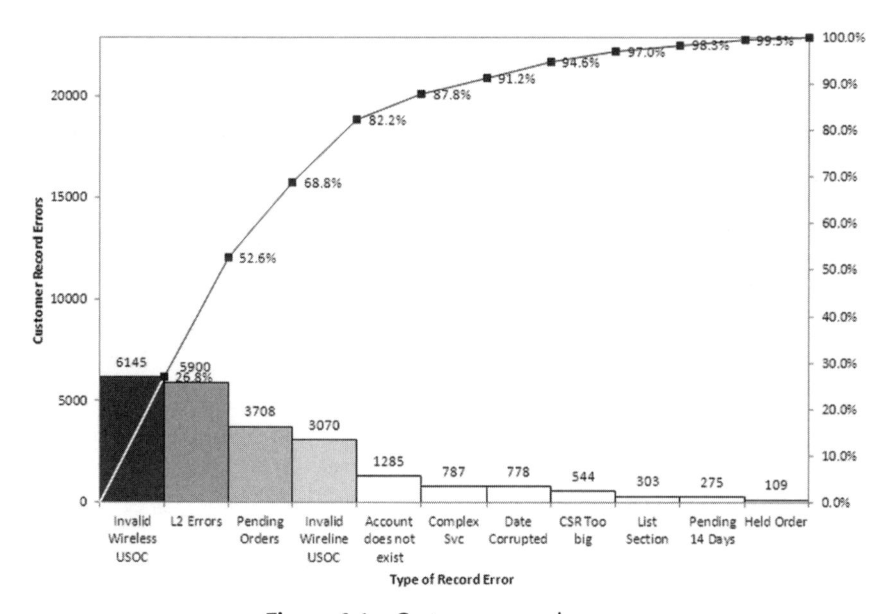

Figure 6.4 Customer record errors.

too different. By restricting ourselves to just one error type per team, we were able to find the root causes with just a half-day meeting per team.

To prepare for the meeting, we printed out 50 to 100 examples of each error because sometimes you discover that you need more examples when you actually start looking at each error. Then:

1. Using all the online systems, we investigated the root cause of each rejected transaction. Again, we restricted ourselves to analyzing just one transaction at a time.
2. As the team agreed on the cause of the rejected transaction, I kept a stroke tally for each root cause (Fig. 6.5). Gradually, as we looked at more and more transactions, a pattern began to reveal itself (e.g., wireless number mismatch). Sometimes it took only 25 transactions; sometimes it took 50. But a pattern began to reveal itself clustered around one or more root causes. The great thing about evaluating transactions one at a time is that it verifies the root causes as you go.
3. Once the team had identified the root causes, we would stop analyzing and spend an hour defining the new requirements. Most of the time, the requirements were too tight, sometimes too loose, and occasionally nonexistent. The systems analyst would then convey these to the programming staff for implementation.

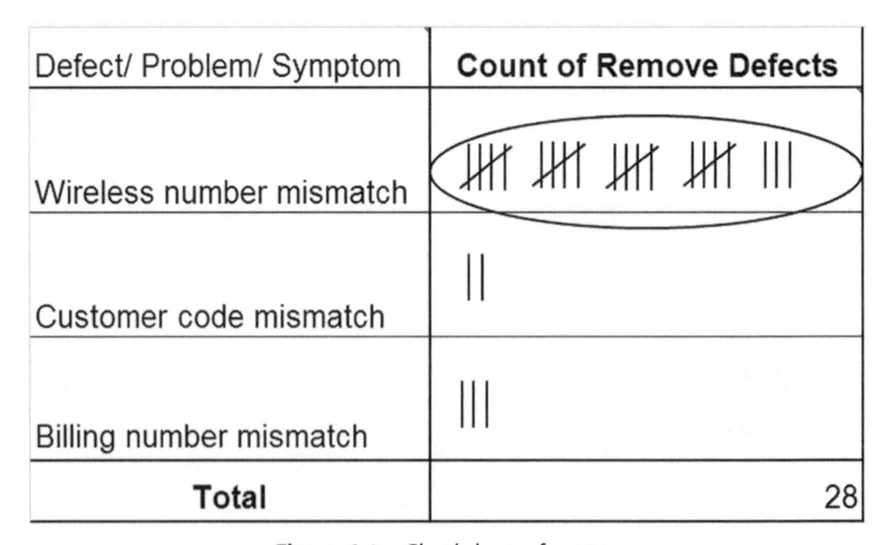

Defect/ Problem/ Symptom	**Count of Remove Defects**
Wireless number mismatch	𝍸 𝍸 𝍸 𝍸 𝍷𝍷𝍷
Customer code mismatch	𝍷𝍷
Billing number mismatch	𝍷𝍷𝍷
Total	28

Figure 6.5 Checksheet of causes.

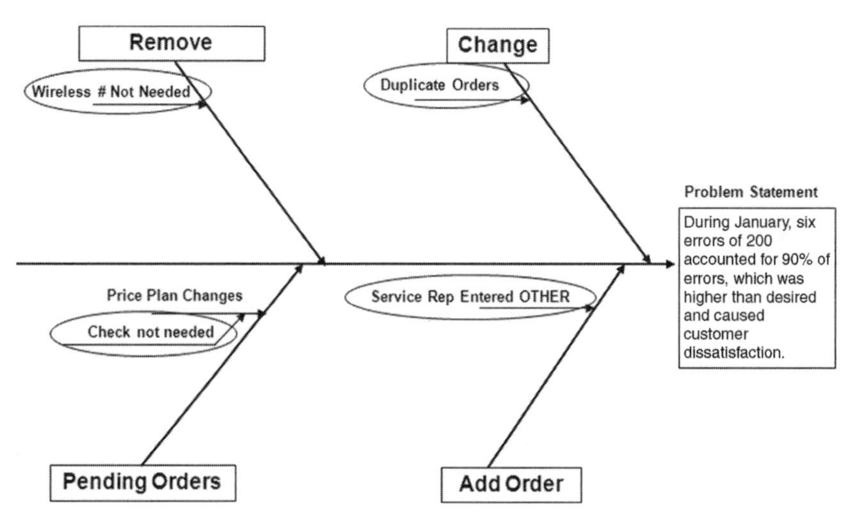

Figure 6.6 Fishbone diagram of root causes.

Once these root causes had been identified, I captured them in an Ishikawa or fishbone diagram (Fig. 6.6).

Analyzing Results

It took four months to implement the revisions, but it was worth it. By midyear, the changes *completely eliminated the two top service-affecting errors* (Fig. 6.7) *and three of the four record-affecting changes* (Fig. 6.8). The process cut total errors from 31,121 down to 2,395 per month—a 77 percent reduction in total errors (Fig. 6.9). This reduction translated to savings of over $3 million per year.

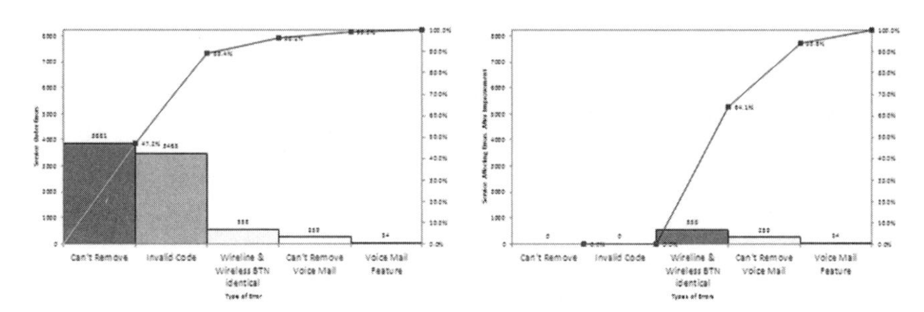

Figure 6.7 Before and after Pareto chart of service-affecting errors.

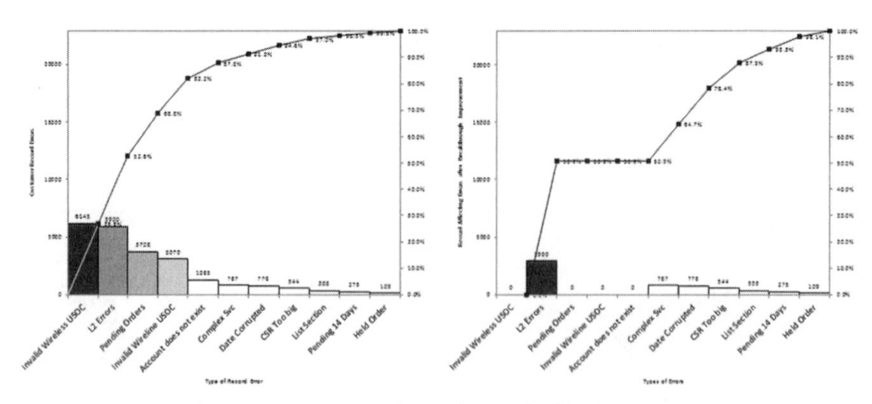

Figure 6.8 Pareto chart of record-affecting errors.

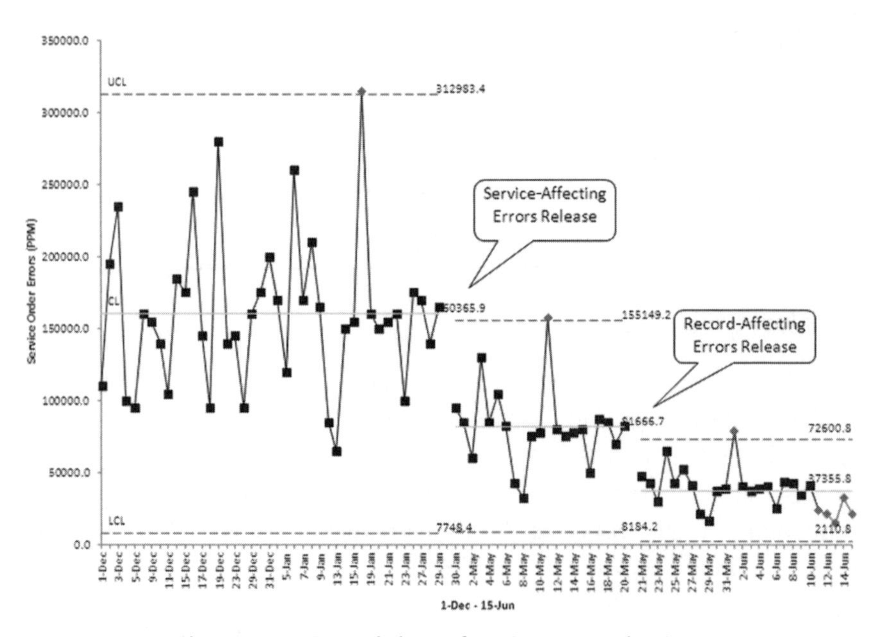

Figure 6.9 Control chart of service error reductions.

Common Problems

Every business has to take orders, process payments, order materials, and pay for them. Every one of these processes suffers from defects, mistakes, and errors. It doesn't matter if you're taking a customer's order for a cell phone or a pizza, if you don't get the order right the first time, it creates problems all the way down the line.

The core elements of any software application involve searching for and then adding, changing, or deleting data. Most applications assume a perfect world, where the data is only created or modified by the system. This is rarely the case. Perfect data continues to be a mythologic assumption that fosters faulty requirements and designs.

The requirements for adding, changing, and deleting data are often too loose, too tight, or nonexistent, which leads to errors and rejected transactions that must be corrected manually by people hunched over computer terminals for eight hours a day (a *fix-it factory*).

Conclusion

Until you get to where you can prevent errors in your processes, every pain point could benefit from a simple, yet rigorous approach to analyzing and eliminating mistakes and errors. The dirty-30 process is ideal because the data required to implement it is collected by most systems automatically. Then all it takes is four hours of analysis to identify the root cause of the error. And you can do it with any error-prone process, not just software.

One of the positive by-products of this approach is that the employees learn firsthand how their actions most often fail. This allows them to learn how to do better the next time.

Fast, Fast, Fast Relief

This process is very simple:

1. *Find the data.* Get a control chart of the pain over time so that you know what you're dealing with (see Fig. 6.1).
2. Once you know the level of pain, *use Pareto charts to drill down* to find the biggest contributors to the pain (see Fig. 6.2). I rarely stop at the first level of Pareto chart. I drill down into the "big bar" to see if there's a Pareto pattern within the big bar (see Fig. 6.3).
3. *Create a problem statement.* The big bar on the Pareto chart becomes the problem statement in the head of the fishbone diagram.
4. *Gather a "SWAT team" of experts* to diagnose the root causes and identify countermeasures. A properly focused root-cause analysis can be done in under four hours.

TAR PIT Most companies make the mistake of convening a team before they know what problem they are trying to solve. By narrowing the focus using data, it's easy to figure out who should be on the team.

5. *Implement countermeasures* to fix the problem and relieve the pain. To verify that the countermeasures actually reduce the pain, add data to the control chart from step 1 (Fig. 6.9). Compare the before and after Pareto charts (Figs. 6.7 and 6.8). *Note:* If the charts don't show improvement, you haven't fixed the problem.
6. *Sustain the improvement* (see Chap. 10). Once the solution has been implemented and verified, just continue to use the control chart (Fig. 6.9) to monitor the cost or count of defects *forever*, and implement corrective actions when it goes out of control.
7. *Repeat until the pain is gone.* This team continued and reduced errors to less than 1 percent the following year. Find a new source of pain and repeat.

The good news is that you can start today; the bad news is that you will never be done. Markets change, competitors change, technology changes, and economies rise and fall. There will always be a new source of pain, but you now have a method for fast, fast, fast relief.

CHAPTER 7

A Picture Is Worth a Thousand Words

In the September 2009 issue of the *Harvard Business Review*, Paul Hemp, in an article entitled, "Death by Information Overload," rants that too much information is flooding our e-mail, Internet searches, business reports, cable news, and so on.

What's the cure for information overload? A well-drawn chart!

One of the ongoing questions I receive from QI Macros users is, "What chart should I choose?"

My friend Dave used to say, "Everything you want to know is written down in a book somewhere. All you have to do is find an author who speaks to you." One of my users recommended a book, *Say It with Charts*, by Gene Zelazny, director of visual communications for McKinsey and Company. He spent 30 years helping McKinsey make sense of the plethora of data in companies. What I admire about this book is its simplicity.

Say It with Charts

Zelazny found that there are five styles and categories of charts (Fig. 7.1).

- *Pie charts*—to show the size of components of the total as a percentage
- *Bar charts*—to show the rank of various items (e.g., a Pareto chart)
- *Column charts*—to show frequency of occurrence (e.g., histograms)
- *Line (i.e., control) charts*—to show time series (i.e., how they change over time)
- *Scatter and tornado charts*—to show correlation (i.e., the relationship between two variables)

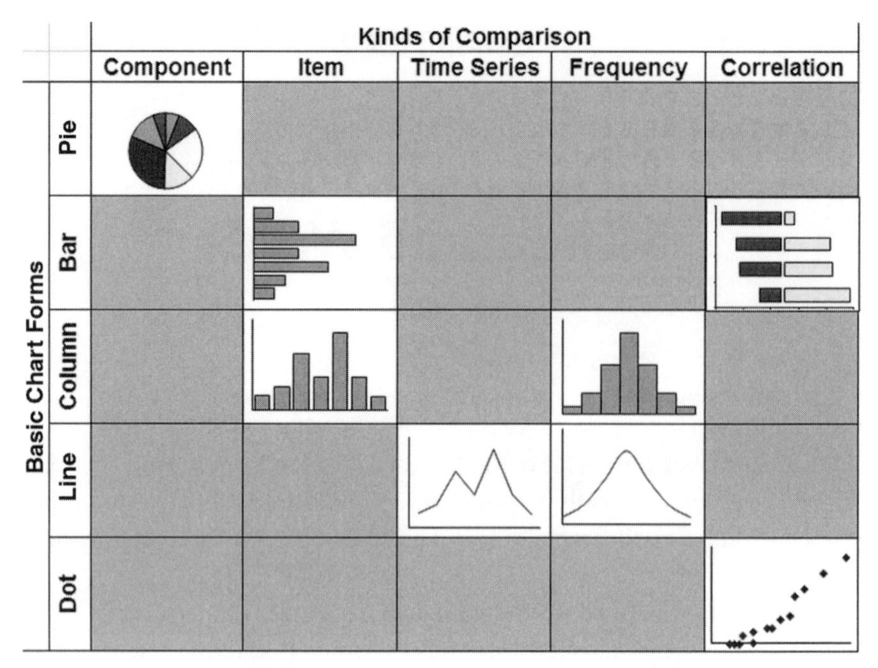

Figure 7.1 Types of charts and applications.

To achieve breakthrough improvement, you will want to combine line and column charts into control charts, Pareto charts, and histograms (Fig. 7.2), but as you can see, these are just combinations of the simple charts in Figure 7.1.

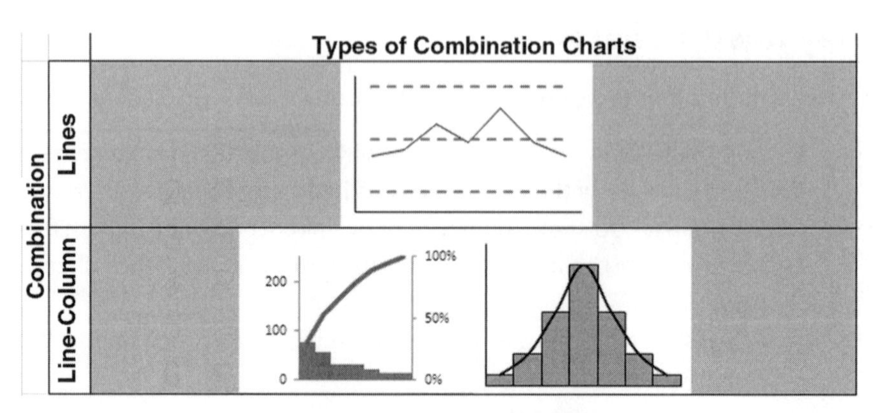

Figure 7.2 Combination charts.

Process

Zelazny recommends a three-step process for choosing a chart.

1. *Message.* What story are you trying to tell? It will shape the choice of chart.
2. *Comparison.* Humans learn by contrast. What are you trying to compare?
3. *Chart selection.* Which chart best delivers your message and comparison?

For example, there can be many ways to show the same data (Fig. 7.3).

Simplify, Simplify, Simplify

Zelazny suggests that instead of trying to put everything onto one chart, we should make several charts with a limited amount of data. My favorite violation of this rule? Comparing your business with that of various competitors. If you put your business and all competitors on one chart (e.g.,

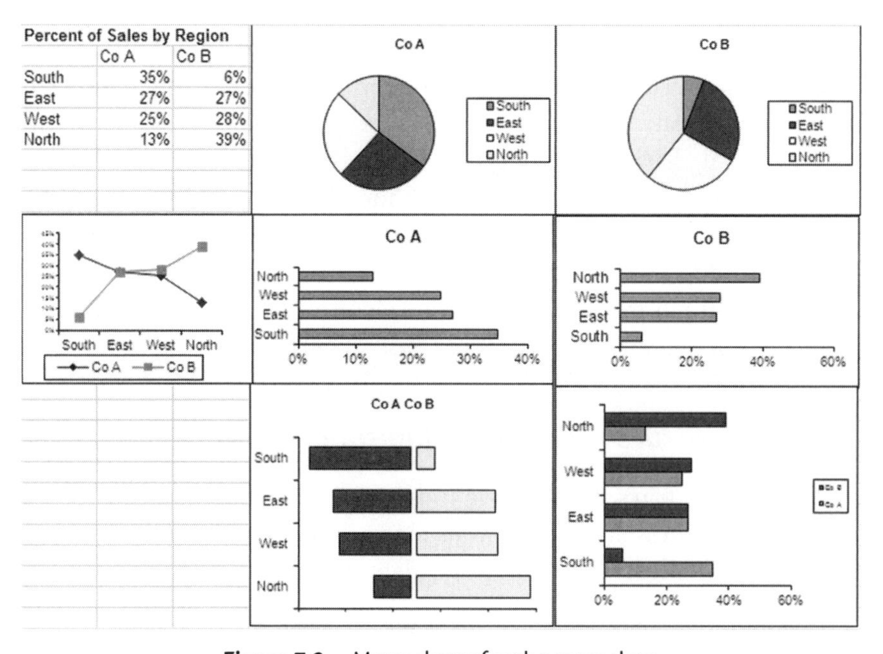

Figure 7.3 Many charts for the same data.

the line chart in Fig. 7.3 or the bar chart in the bottom right-hand corner), it becomes difficult to determine where you stand compared with everyone else.

The bar charts in the middle of Figure 7.3 split Company A into one chart and Company B into another chart. From these two charts, it's easy to see that Company A is big in the South, whereas Company B is big in the North. The tornado or paired bar chart at the bottom left of Fig. 7.3 does a good job of showing this on one chart. It isn't a standard chart in Excel, but it is available in the QI Macros.

Split data out onto multiple charts comparing your business with that of one competitor per chart . . . and you quickly discover which competitors you need to watch.

Chart Junk

I recently stumbled onto the book, *Visual Explanations*, by Edward R. Tufte (Graphics Press, Cheshire, CT, 1997). The *New York Times* calls Tufte "the Leonardo da Vinci of data." The author says that there are right ways and wrong ways to show data; there are displays that reveal the truth and displays that do not.

Having exhibited at various health care and manufacturing conferences, I've seen hundreds of improvement projects displayed around the exhibit hall. Few use charts; most just use words. The charts used are often incorrect for the type of data being shown.

The right charts and data can tell an improvement story quickly and easily. Words take too long. In 2010, about 60 percent of the Baldridge Award applicants were health care companies. One of the things a Baldridge examiner told me is that after looking at hundreds of applications, most improvement stories show the before picture, but few show the *after-improvement results using control and Pareto charts.*

The Right Picture Is Worth a Thousand Words

Information displays should serve the analytical purpose at hand. This is why I use the QI Macros to draw as many different charts as possible to explore which one tells the best story. Here are some of Tufte's insights:

- *Numbers become evidence by being in relation to something* (i.e., comparison).

- *The disappearing legend.* When the legend on a chart is lost, the insights can be lost as well.
- *Chart junk.* Good design brings absolute attention to data. Bad design loses the insights in the clutter.
- *Lack of clarity in depicting cause and effect.*
- *Wrong order.* A fatal flaw can be in ordering the data. A time series (i.e., a control chart) may not reveal what a bar chart (i.e., a histogram) might reveal.

I usually draw as many different charts from the same data as I can to see which one tells the best story. You should too. Every picture tells a story, but some pictures are better than others at telling the story. The QI Macros make it easy to draw one chart after another so that you can quickly discard some of them and select others that engage the eye in the real issues.

As Tufte would say, "Don't let your charts become *disinformation*." There's enough of that in the world already.

Chart junk is a form of disinformation. It confuses the reader. Clean up your charts. Get rid of unnecessary clutter. Choose the right kind of chart for your data, and you'll go a long way toward motivating your readers to understand and align with the business case presented.

QI Macros Chart Wizard

If you're not sure what chart to choose for your data, select it with your mouse, and click on the QI Macros "Chart Wizard" (Fig. 7.4).

The QI Macros Chart Wizard will analyze your data and choose the right charts for you. It saves you time and mistake-proofs the chart selection process. Spend your time on more important things, such as analyzing the charts after they are drawn.

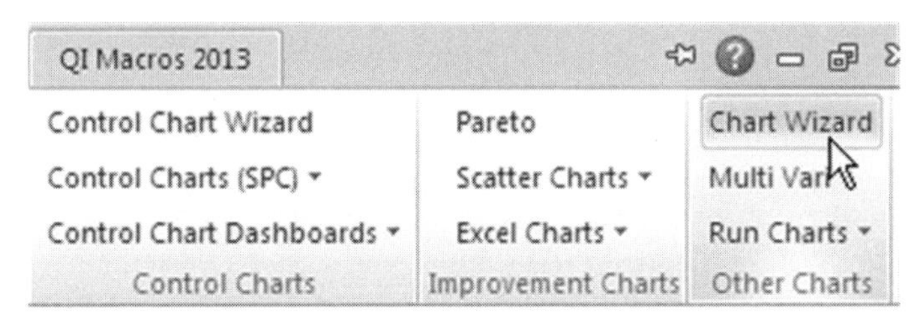

Figure 7.4 QI Macros Chart Wizard.

To use the QI Macros Chart Wizard:

1. Just select your data.
2. Then select "Chart Wizard" from the QI Macros menu.
3. The Chart Wizard will draw one or many charts based on your data.

Example

1. Select your data (Fig. 7.5)—QI Macros Test Data\histogrm.xls.

	A	B	C	D	E	F	G	H
1	Sample	Obs 1	Obs 2	Obs 3	Obs 4	Bursting Strength of Bottles		
2	S1	265	205	263	307	220		
3	S2	268	260	234	299	215		
4	S3	197	286	274	243	231		
5	S4	267	281	265	214	318		
6	S5	346	317	242	258	276	USL = 346	
7	S6	300	208	187	264	271	LSL = 200	
8	S7	280	242	260	321	228		
9	S8	250	299	258	267	293		
10	S9	265	254	281	294	223		

Figure 7.5 Data for charts.

2. Click on "Chart Wizard," and the wizard will draw all the most probable charts and run "Descriptive Statistics" (Fig. 7.6). Descriptive Statistics gives histograms, box plots, confidence intervals, and normal-probability plots.

Because there is more than one sample, the Chart Wizard will draw box and whisker plots of the data by row (Fig. 7.7) and column (Fig. 7.8).

Because there are five samples, the Chart Wizard will draw an XbarR chart (Fig. 7.9).

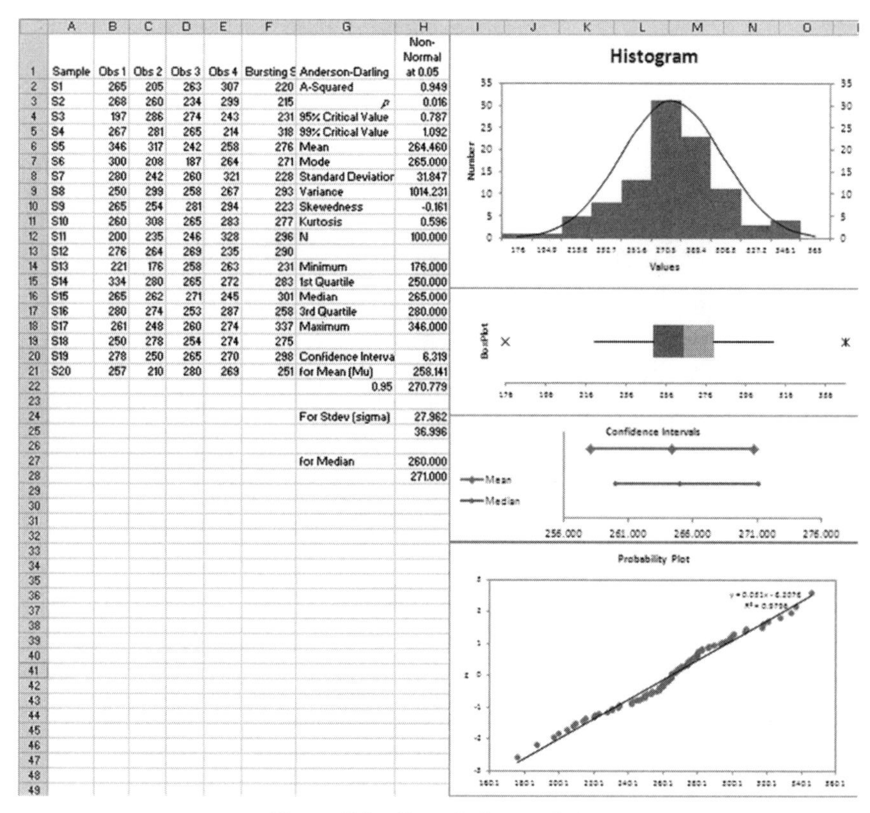

	A	B	C	D	E	F	G	H
1	Sample	Obs 1	Obs 2	Obs 3	Obs 4	Bursting S	Anderson-Darling	Non-Normal at 0.05
2	S1	265	205	263	307	220	A-Squared	0.949
3	S2	268	260	234	299	215	p	0.016
4	S3	197	286	274	243	231	95% Critical Value	0.787
5	S4	267	281	265	214	318	99% Critical Value	1.092
6	S5	346	317	242	258	276	Mean	264.460
7	S6	300	208	187	264	271	Mode	265.000
8	S7	280	242	260	321	228	Standard Deviatior	31.847
9	S8	250	299	258	267	293	Variance	1014.231
10	S9	265	254	281	294	223	Skewedness	-0.161
11	S10	260	308	265	283	277	Kurtosis	0.596
12	S11	200	235	246	328	296	N	100.000
13	S12	276	264	269	235	290		
14	S13	221	176	258	263	231	Minimum	176.000
15	S14	334	280	265	272	283	1st Quartile	250.000
16	S15	265	262	271	245	301	Median	265.000
17	S16	280	274	253	287	258	3rd Quartile	280.000
18	S17	261	248	260	274	337	Maximum	346.000
19	S18	250	278	254	274	275		
20	S19	278	250	265	270	298	Confidence Interva	6.319
21	S20	257	210	280	269	251	for Mean (Mu)	258.141
22							0.95	270.779
23								
24							For Stdev (sigma)	27.962
25								36.996
26								
27							for Median	260.000
28								271.000

Figure 7.6 Descriptive statistics.

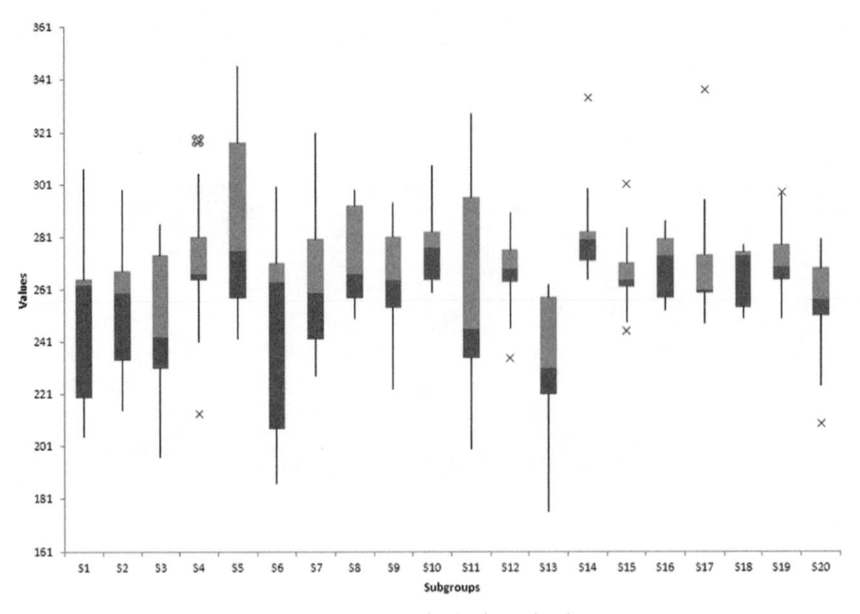

Figure 7.7 Box and whisker plot by row.

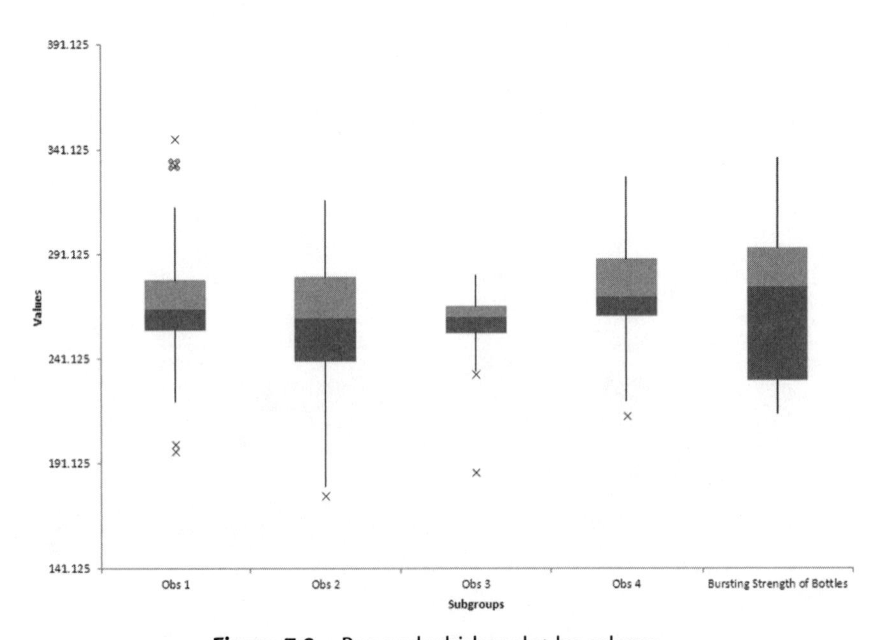

Figure 7.8 Box and whisker plot by column.

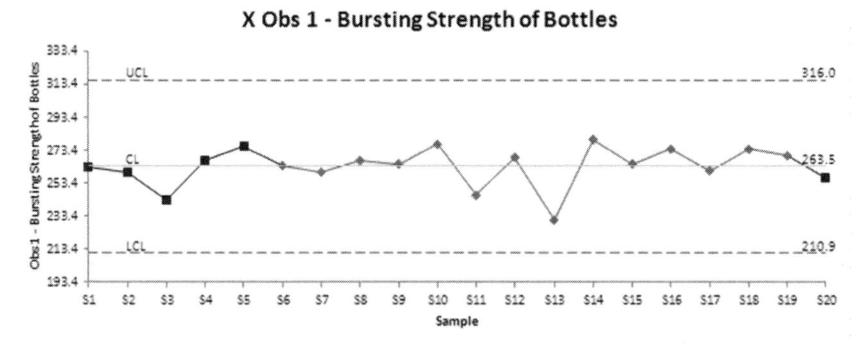

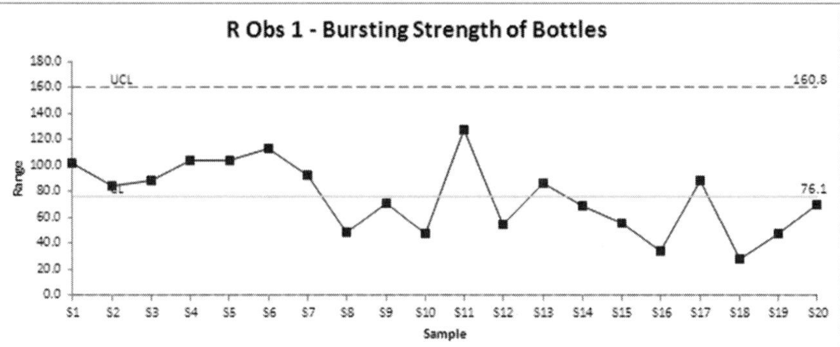

Figure 7.9 XbarR chart.

Here's how the QI Macros Chart Wizard selects the chart:

Columns of Data	1	2	3–9	10+
Integers or decimals	If Max > 10*Min and row headings are not dates: Pareto chart			
Integers	*c* Chart	If numerator/ denominator: *p* and *u* Chart Scatter If denominator is constant: *np* Chart		
Integers or decimal	Descriptive Statistics Includes histogram, normal probability plot, box plot, and confidence intervals			
		Box and whisker by row and column		
Normal	XmR chart	XbarR chart		XbarS chart
Nonnormal	XmR median chart	XmedianR chart		XbarS chart

If you're not sure what chart to choose, the Chart Wizard will help get you started. Then you can refine your choices from there.

Here's My Point

A tsunami of data threatens to overwhelm all of us. Consolidating that data into a handful of well-designed charts can quickly communicate performance trends in ways that no spreadsheet ever will.

CHAPTER 8

Excel Power Tools for Breakthrough Improvement

Breakthrough improvement thrives on charts, graphs, and diagrams of performance data. To succeed at breakthrough improvement, you'll need a set of power tools to analyze and graph your data.

When I learned quality improvement back in 1989, I had to draw all the charts by hand. I spent five days in a control-chart class calculating all the formulas using a handheld calculator and drawing the various charts. Most of my fellow students and I struggled to calculate the formulas correctly. At the end of the course, we had only spent two hours on what the charts were telling us. I knew that there was no way I was going to get phone company personnel to draw control charts by hand, but I couldn't get my management to spend $1,000 on Statistical Process Control (SPC) software, so I just struggled along.

After I left the phone company in 1995, I started experimenting with using Excel to draw the charts necessary for breakthrough improvement. I launched the first, primitive version of the QI Macros in 1997 and have been improving them ever since. Because Excel does all the heavy lifting—drawing the charts—I can keep the cost low for the typical user.

Microsoft Excel is a tremendously powerful tool for breakthrough improvement, but most people don't even know how to use the basic capabilities of Excel. If you think you're a hotshot Excel user, read on because I'll show you how to use the QI Macros for Excel. If you don't own a copy of Excel or Office, you can usually pick up inexpensive copies of older versions at ebay.com. The QI Macros work in all versions of Excel. Note: Microsoft will drop support of Excel 2003 and Windows XP in April of 2014. As of the publication of this book, Excel 2010 is the most reliable new version. Excel 2013 still crashes too often.

Setting Up Your Data in Excel

How you set up the data can make analysis easy or hard. Using an Excel *worksheet*, you can create the labels and data points for any chart—control chart, Pareto chart, histogram, or scatter plot. This gives you a worksheet that looks like Figure 8.1.

Prepare Your Data

Other breakthrough improvement software packages make you transfer your Excel data into special tables, but not the QI Macros. Just put your data in a standard Excel worksheet. The simplest format for your data is usually one column of labels and one or more columns of data, but it can also be in rows. (Once you've installed the QI Macros, see Documents\QI Macros Test Data for sample data for each chart.)

Once you have your data in the spreadsheet, you will want to select it to be able to create a chart. Using your mouse, just highlight the data to be graphed (i.e., select by clicking the mouse button and dragging it up or down), run the appropriate QI Macros chart, and Excel will do the math and draw the graph.

	A	B	C	D	E
1	Batch Number	Viscosity		Month	Falls/1000 Patient Days
2	B1	33.75		06/28/09	3.6
3	B2	33.05		07/29/09	4.5
4	B3	34.00		08/29/09	4.7
5	B4	33.81		09/28/09	6.0
6	B5	33.46		10/29/09	4.6
7	B6	34.02		11/28/09	3.6
8	B7	33.68		12/29/09	7.6
9	B8	33.27		01/29/10	7.7
10	B9	33.49		02/26/10	5.6
11	B10	33.20		03/29/10	5.7
12	B11	33.62		04/28/10	7.0

Figure 8.1 Formatting data in Excel.

Tips for Selecting Your Data

- Click and drag with the mouse to select the data.
- To highlight cells from different columns (Fig. 8.2), click on the top-left cell, and drag the mouse down to include the cells in the first row or column. Then hold down the Control key (CTRL) while selecting the additional rows or columns.
- You also may use data in horizontal rows (Fig. 8.3), but it's not a good format for data in Excel. Whenever possible, put your data in columns, not rows.
- Excel formats most numbers as "General" not "Number." If you do not specify the format for your data, Excel will choose one for you. To get desired precision, select the data with your mouse, choose "Format–Cells–Number," and specify the number of decimals.
- Don't select the entire column (65,000+ data points) or row, just the cells that contain the data and associated labels you want to graph.

	A	B	C	D	E
1	Month	Falls/1000 Patient Days		Total Patient Falls	Total Patient Days
2	Jan-04	3.6		17	4658
3	Feb-04	4.5		22	4909
4	Mar-04	4.7		23	4886
5	Apr-04	6.0		30	4970
6	May-04	4.6		22	4780
7	Jun-04	3.6		18	4973
8	Jul-04	7.6		44	5762
9	Aug-04	7.7		42	5441
10	Sep-04	5.6		33	5893

Figure 8.2 Selecting non-adjacent cells.

	A	B	C	D	E	F	G	H	I	J	K	L	M
1	Month	J	F	M	A	M	J	J	A	S	O	N	D
2	Patient Satisfaction %	82	79	84	82	92	80	94	78	83	84	92	84

Figure 8.3 Horizontal data.

- When you select the data you want to graph, you can select the associated labels as well (e.g., Jan, Feb, Mar). The QI Macros will use the labels to create part of your chart (e.g., title, axis name, legend).
- Data should be formatted as numbers. Your data must be numerical and formatted as a number for the macros to perform the necessary calculations. Data exported from Microsoft Access, for example, is exported as text. If you have numbers formatted as text, the QI Macros will attempt to convert them to numbers before drawing the chart.
- Select the right number of columns. Each chart requires a certain number of columns of data to run properly. They are:
    - *One column:* Pareto, pie, *c* chart, *np* chart, XmR chart
    - *One or more columns:* line, run, bar, column, pie, histogram
    - *Two columns:* scatter, *u* chart, *p* chart
    - *Two or more columns:* box and whisker, multivari, XbarR, XbarS
- Beware of hidden rows or columns. If you select columns A through F but columns B and C are hidden, the QI Macros will use all five columns, including the hidden ones. To select nonadjacent columns, use the Control key (CTRL).

Data Collection and Measurement

While most companies have too much data, people can always identify something they aren't tracking that they should be tracking. Then they think that they have to set up a whole system to collect the measurement. This is a mistake. You don't know if the measurement is useful until you have collected it for awhile. Rather than wait for an automated measurement system, start today using a few simple tools: a check sheet or a log of errors.

I use these kinds of check sheets when I'm working with a team on the Dirty-30 process for breakthrough improvement. They find causes, and I write them down and tally the number of times each occurs. By the thirtieth data point, a Pareto pattern appears that points us at the most common (i.e., root) cause of the problem.

Check Sheets for Data Collection

Nothing could be simpler than data collection with a check sheet. The QI Macros have a template in the improvement tools to get you started (Fig. 8.4). Simply print it out and start writing on it. What to write?

	A	B	C	D	E	F	G	H
1					Week			
2	Defect/ Problem/ Symptom	M	Tu	W	Th	F	Sa	Total
3	Delay							0
4	Missed Commitments							0
5								0
6	Defects							0
7	Errors							0
8								0
9	Repeat Fixes							0
10								0
11	Total	0	0	0	0	0	0	0

Figure 8.4 QI Macros check sheet.

In column A, write the first instance of any defect, problem, or symptom you detect. For example, if someone is calling you for support and they have a problem with duplicate date of service, then put "duplicate date of service" in A3 and put a stroke tally in the day of the week (e.g., Monday). Then continue adding to the check sheet as the week goes on, adding defects, problems, or symptoms. By the end of the week, you'll have an interesting picture of support calls (Fig. 8.5).

Just add up the number of calls and one bucket or another will jump out as the majority of the calls. *Hint:* Use a Pareto chart (Fig. 8.6) to show most common support calls.

Get the idea? Use a check sheet to prototype your data-collection efforts. Iterate until you start to understand what you really need to know to make improvements. A series of check sheets may be all you need to solve a pressing problem. If necessary, you can implement a measurement system to collect the data over time.

	A	B	C	D
1				
2	Defect/ Problem/ Symptom	M	Tu	W
3	Duplicate Date of Service	卌 |||	卌 卌 |	卌 卌
4	No Auth	|||	||	||||

Figure 8.5 Check sheet of rejected claims.

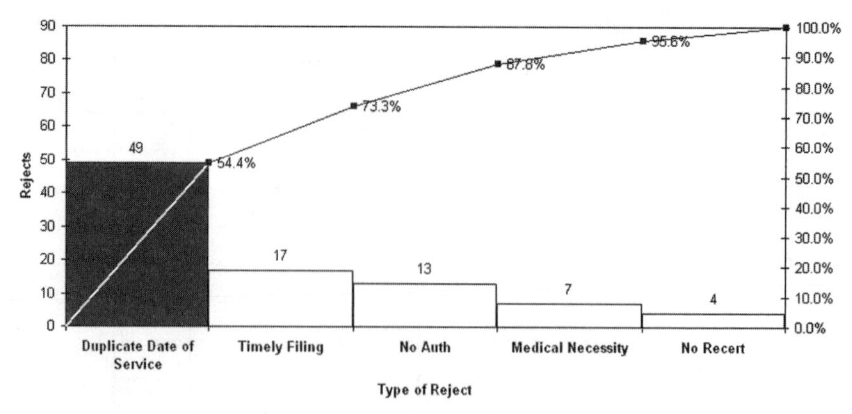

Figure 8.6 Pareto chart of rejects check sheet.

So please don't wait for a magical, all-encompassing measurement system to deliver data. It's not going to happen. And I often find that this is just an excuse to avoid making improvements ("I can't because I don't have the measurements I need").

Haven't you waited long enough to start making measurable improvements (even if your data-collection tool is just a simple check sheet)? Or are you going to keep letting loads of cash slip through your fingers? All it takes is a check sheet and a pencil. Get on with it.

The QI Macros for Excel

The QI Macros breakthrough improvement software consists of five main parts (Table 8.1).

- Over 40 tools to draw control charts, histograms, Pareto charts, and so on. The Chart Wizard and Control Chart Wizard will automatically choose the right chart for you.
- Over 90 fill-in-the-blank templates of breakthrough improvement forms, tools, and charts.
- Three fill-in-the-blank dashboard tools for XmR, XbarR, *c*, *np*, *p*, and *u* charts.
- Statistical tools such as ANOVA, *t* test, regression, and so on. The Stat Wizard will automatically run all possible statistics on the data.
- Data-transformation tools such as the PivotTable Wizard, Word Count, Stack, and Restack.

Table 8.1 QI Macros Components

1. Macros	2. Templates	3. Statistics	4. Data Transformation	5. Control Chart Dashboards
Control charts	Control charts	ANOVA	PivotTable Wizard	*c*, *np*, *p*, and *u*
Histograms	Flowcharts	Regression	Word count	XmR
Line, run, or scatter charts	Fishbones	Sample size	Stack/restack	XbarR
Pareto, bar, or pie charts	Gage R&R	*t*-test, *F*-test		
Box and whisker diagrams	DOE and QFD	Chi-squared test		
Multivari charts	FMEA and PPAP	Correlation		

The QI Macros Are Easy to Use

Because I'd never been exposed to software developed before the now-familiar point-and-click mouse-driven interface, I was free to think outside the interface design imposed by minicomputers. I took a "grab-it-and-go" approach to the software—select data with the mouse, and then click on a menu to draw a chart.

The QI Macros is the only data-analysis software that asks you to select your data *before* you choose a type of chart or statistic. Because of this simple design choice, the QI Macros can help you select the right chart or statistic every time *without days of training.*

The QI Macros are easy to learn and use:

- Because the software was developed from the ground up to work in Excel and deliver immediate results in business environments using "grab-it-and-go" simplicity.
  - *Mistake-proof selection of data* (Your data can be in connected or separated rows or columns; the QI Macros will clean up non-numerical data, fix any misalignments, and use your data *as you selected it.*)
  - *Control Chart Wizard* to select the right control chart for you automatically

  - *Control chart dashboards* to simplify monthly reporting
  - *PivotTable Wizard* to simplify analyzing complex transaction files
  - *Stat Wizard* to simplify statistical analysis
  - *Mistake-proof statistical analysis* (Excel can be picky and even produce invalid results if the data isn't used correctly.)
- Because the QI Macros do all the math and statistics for you. (There are no complex formulas to grasp or apply, just charts and results.)
- Because of the fill-in-the-blanks, paint-by-numbers simplicity of the 90+ chart and documentation templates.
- Because you don't have to waste time transposing or transferring your data from Excel to a separate program. (The QI Macros work inside Excel.)

QI Macros Introduction

Many kinds of graphs, forms, and tools are used in breakthrough improvement. Ninety percent of common problems can be diagnosed with control charts, histograms, Pareto charts, and Ishikawa diagrams. Control charts will help you sustain the improvements. The QI Macros and Excel can be used to create all these charts, graphs, forms, and tools.

Installing the QI Macros

To install the QI Macros, simply:

1. Go to my website www.qimacros.com/breakthrough, and fill in your e-mail address to download the QI Macros and the other free breakthrough improvement quick reference cards. This will also sign you up for the free QI Macros lessons online.
2. Download the QI Macros 90-day trial copy (PC or Mac) by clicking on the icon.
3. Double-click on "QIMacros90day.exe" (PC) or "QIMacros90Day.dmg" (Mac) to install the QI Macros.
4. When you start Excel, the QI Macros menu will appear on the ribbon menu in Excel 2007–2013 or Excel's toolbar in Excel 2000–2003 and Excel 2011 for Mac.
5. If you have any problems, check my website www.qimacros.com/support.

Sample Test Data

The QI Macros for Excel install test data on your PC in "Documents/QI Macros Test Data." Use this data to practice with the charts and to determine the best way to format the data before you run a macro.

Creating a Chart Using the QI Macros Menu

There are two different ways to create charts in the QI Macros.

1. Select your data, and then run a chart from the menu. To create a chart using the menu, just select the data with your mouse. Then, using the QI Macros menu (Fig. 8.7), select the chart you want to create. The QI Macros will do the math and draw the graph for you.
2. Use the fill-in-the-blanks chart templates.

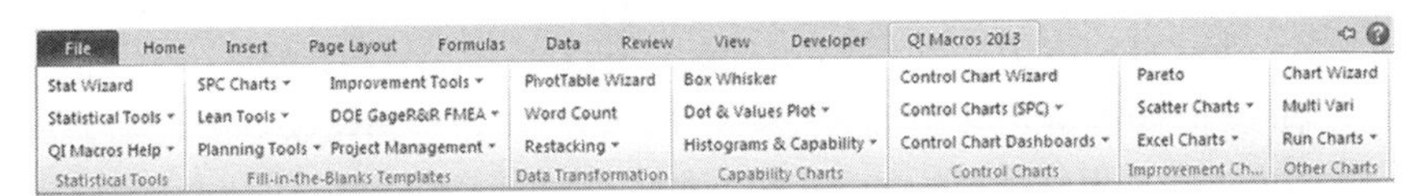

Figure 8.7 QI Macros menu.

Create a Control Chart

1. *Open a workbook* (e.g., AIAGSPC.xls or Healthcare SPC.xls, which has worksheets for the common control charts and Pareto charts).
2. *Select the labels and data to be graphed* (e.g., XmR data–Falls per 1,000 patient days). Click on the top-left cell, and drag the mouse across and down to include the cells on the right.

TIP Use Excel's shortcut keys. I have found that the CTRL-SHIFT keys combined with the arrow keys, END, and HOME can speed up data selection.

- CTRL-SHIFT-ARROW key (up, down, left, or right) selects cells to the last nonblank cell in the same column or row.
- CTRL-SHIFT-END selects cells to the last used cell on the worksheet.
- CTRL-SHIFT-HOME selects cells to the beginning of the worksheet.

---

3. *From the QI Macro menu, select "Control Chart Wizard."* Excel will start drawing the graph. Fill in the graph title and the *X*- and *Y*-axis titles as appropriate.
4. *To add text to any part of the graph,* just click anywhere on the white space and type. Then use the mouse to click and drag the text to the desired location. *To change titles or labels,* just click and change them. Change other text in the worksheet in the same way.
5. *To change the scale on any axis, double-click on the axis.* Select "Scale," and enter the new minimum, maximum, and tick-mark increments.
6. *To change the color of any part of the graph, double-click on the item to be changed.* A patterns window will appear (Fig. 8.8). Select "Font" to change text colors, "Line" to change line colors and patterns, or "Marker" to change foreground and background colors. Line graphs showing defects or delay are the key first step of any problem solution.
7. *To change the style of any line on the graph,* double-click on the line. The window (Fig. 8.9) is displayed. Changing the line style, color, and weight is all performed in this window. When you're done, click "OK." The changed graph is now easier to read.
8. *To change the style of graph, right-click on the chart, and choose "Chart Type."* Click on the desired graph format and then "OK."

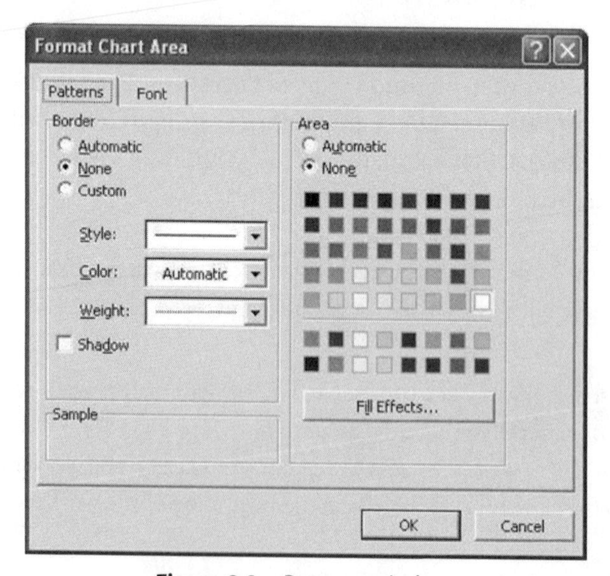

Figure 8.8 Patterns window.

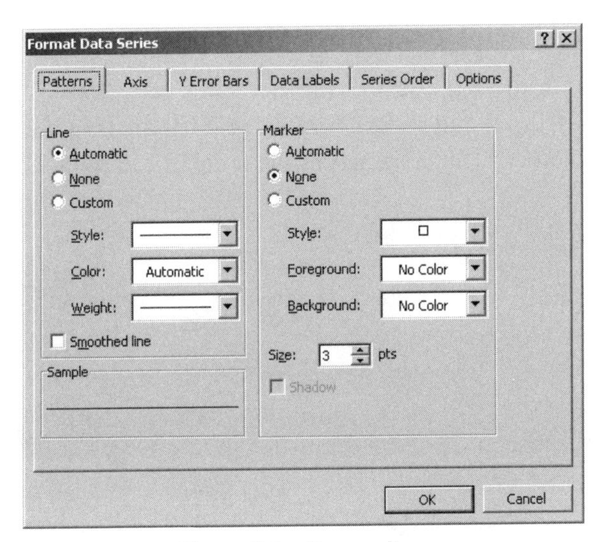

Figure 8.9 Format line.

How to Use Custom Excel Chart Formats

Several QI Macros customers have said, "I'd like to use a different format for some of the charts I produce. I don't want to have to redo it all manually every time." *Good news!* In Excel 2007–2013 you can format just about any chart and then *save the format as a template that you can apply to any future chart.*

Saving a Reusable Chart Template

Once you've drawn a chart, you can change the formatting of the lines, fonts, colors, whatever you want. In Figure 8.9, I've changed a *c* chart to use Broadway font and other line styles.

If I then click on "Chart Tools–Design–Save As Template" (Fig. 8.10) and give it a name such as "*c* Chart," I can then access this format later.

Then I can reuse this template every time I draw a *c* chart.

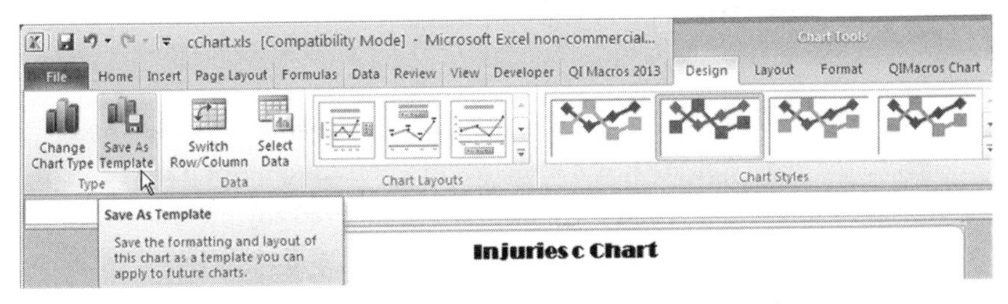

Figure 8.10 Save a custom chart format as template.

Applying the Template to a New Chart

Thus, if I draw another *c* chart and then click on "Chart Tools–Design–Change Chart Type" and select "Templates–*c* Chart "(Fig. 8.11), Excel will reapply all the formatting changes in the template. *Pretty slick, huh?*

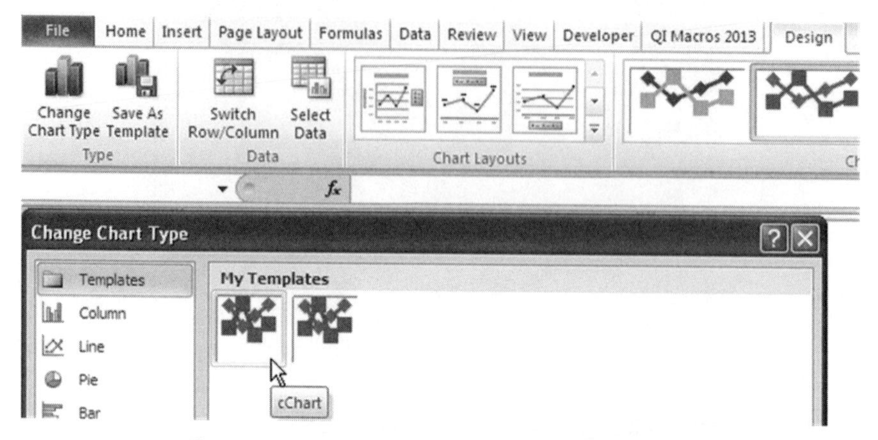

Figure 8.11 Change chart type to custom format.

WARNINGS This should work for most of the charts except those that use error bars, such as:

- XbarR, XbarS, XmedianR
- *P* and *u* charts
- Box and whisker plots

WORKAROUND If you switch the control chart format to "Wave," the *X*, *p*, and *u* charts will use normal formatting that can be modified by this technique. Click on "Control Charts (SPC)–Control Chart Rules–p/u UCL/LCL Format" (Fig. 8.12) to get the "Wave" format.

TAR PITS I found out the hard way that *Excel will remember everything about the chart format!* It will remember red out-of-control points and lines on control charts and change those points to red anytime you apply the format. If the chart has fixed *X*- or *Y*-axis upper and lower limits, Excel will remember those too. So make sure that the axis options are set to

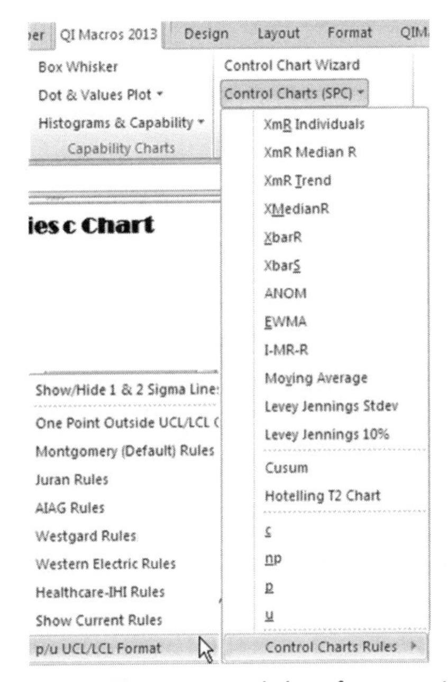

Figure 8.12 Change control chart format to Wave.

auto, not a fixed value. Otherwise, when you apply the format, your chart may vanish because the data is below the fixed limit. Thus you might want to use QI Macros "Chart Menu–Clear Stability Analysis" *before saving the template*. I haven't tried this with every QI Macros chart or template to see if it works, but if it works for your charts, it might save you a lot of time.

In November, 2013, I added tools to the QI Macros Chart menu to automate remembering and applying chart formats. Once you've created a chart and changed the format to your liking, click on the QI Macros Chart menu and select "Remember Format." The next time you create a similar chart, you can use "Apply Format" to apply all of the "remembered" formatting changes.

Fill-in-the-Blanks Templates

In addition to the charts listed on the menu, the QI Macros contain over 90 fill-in-the-blank templates. To access these templates, select the "Fill-in-the-Blanks" templates on the QI Macros menu (Fig. 8.13).

SPC Charts ▾ Improvement Tools ▾
Lean Tools ▾ DOE GageR&R FMEA ▾
Planning Tools ▾ Project Management ▾
Fill-in-the-Blanks Templates

Figure 8.13 Fill-in-the-Blanks templates.

Each template is designed for ease of use. Tools such as the flowchart and fishbone diagram make use of Excel's drawing toolbar (see Chap. 11). To view Excel's "Drawing Toolbar," select "Insert Shapes" in Excel 2007–2013 or "View–Toolbars–Drawing" in Excel 2000–2003.

Creating a Control Chart with a QI Macros Template

To monitor and control performance, companies will want to add data to their charts every month. To simplify this process, the QI Macros contain templates for each kind of control chart. Just cut and paste or input data directly into the yellow area. The control charts will populate as you input the data.

These templates are especially helpful if you have novice personnel who will be inputting data or you don't have enough data to run a macro (you're

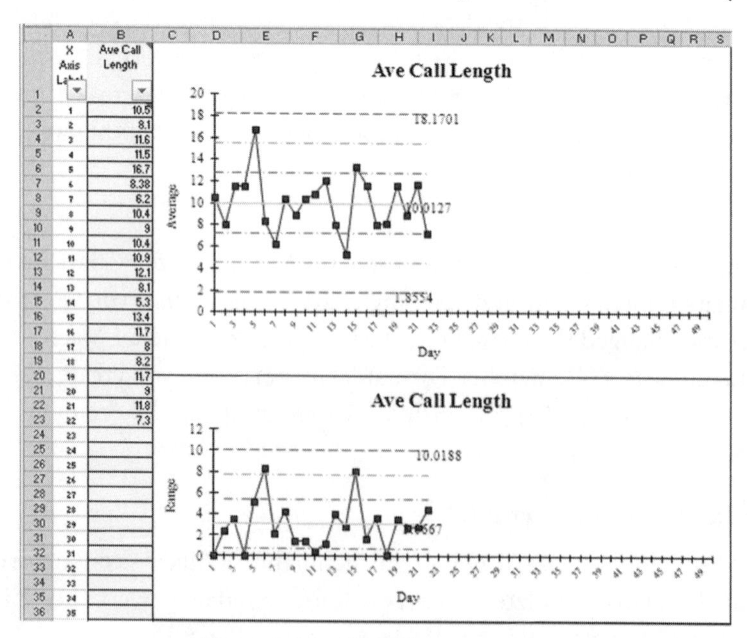

Figure 8.14 XmR chart template.

just starting to collect the data). To create a chart using a template, click on the QI Macros menu, and select the "Fill-in-the-Blanks" templates. Click on the template you want to use (e.g., "SPC Charts–XmR Five Pack").

The input areas for most of the templates start in column A (Fig. 8.14). Either input your data directly into the yellow cells on the template or cut and paste it from another Excel spreadsheet. As you input data, the chart will populate to the right. The *X*-chart templates also display a histogram, probability plot, and capability plot (Fig. 8.15).

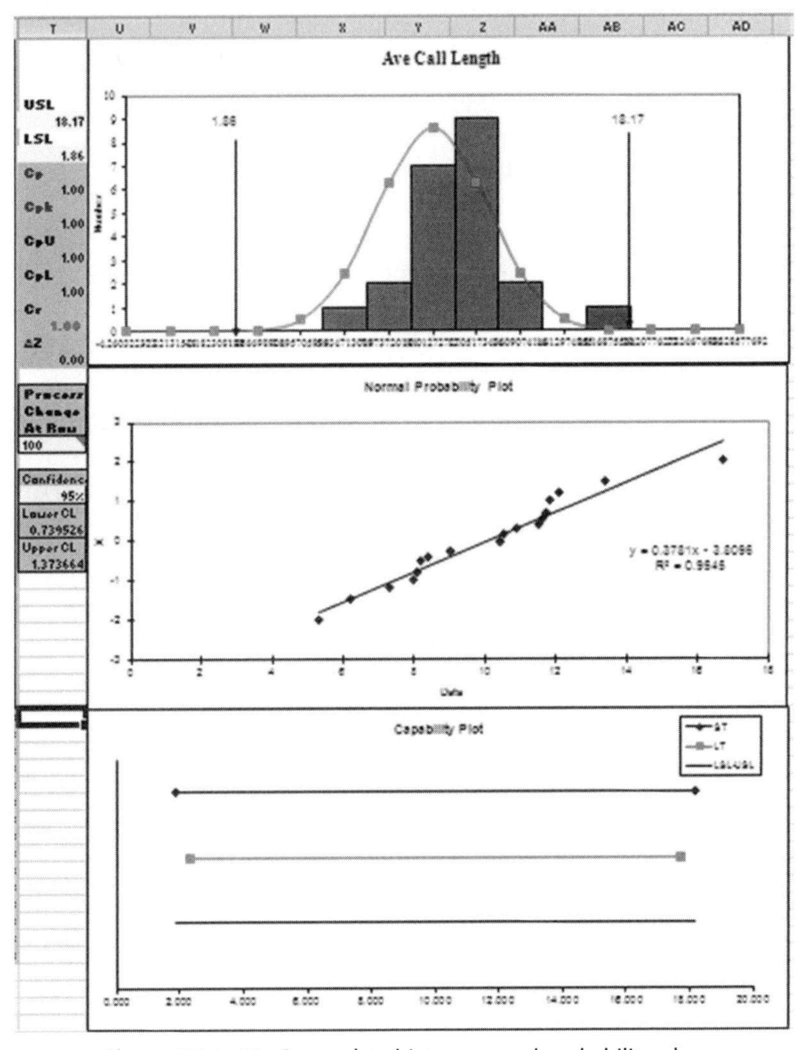

Figure 8.15 XmR template histogram and probability plot.

Running Stability Analysis on a Chart Created by a Template

To run stability analysis on a chart created using a control chart template, just click on the chart (dark boxes will appear at the corners), click on the QI Macros "Chart menu" (Fig. 8.16), and select "Analyze Stability."

Figure 8.16 QI Macros Chart menu.

Choosing Which Points to Plot

Each template defaults to 50 data points. If you have fewer than 50 data points and only want to show the points with data, click on the arrow in cell B1. This will bring up a menu (Fig. 8.17). Uncheck "Blanks" to plot only the points with data.

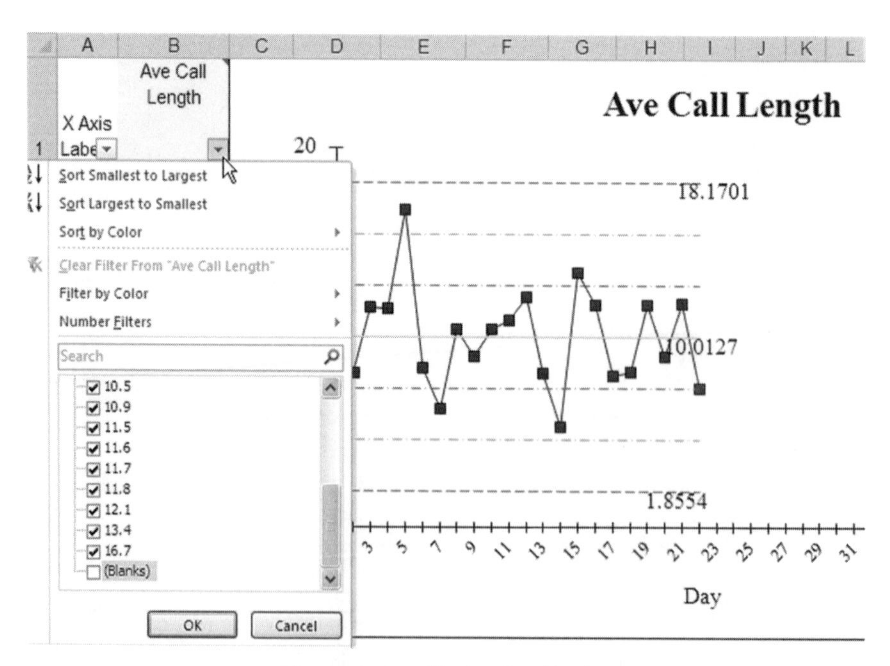

Figure 8.17 Eliminating blanks.

In addition to control charts, there are templates for histograms with Cp and Cpk, precontrol charts, probability plots, Pareto charts, and many more.

The QI Macros "Control Chart Dashboards" have a data worksheet for up to 120 measurements (www.qimacros.com/control-chart/xmr-control-chart-dashboard). Once the data is entered, a single click of the "Create Dashboard" button will create a single sheet containing all of the charts. Each month, just add data to the worksheet, and press "Update Charts" to update all of the charts.

Templates for Your Improvement Efforts

Examples of other templates you will find in the QI Macros include

- *Focus your improvement efforts* using the Balanced Scorecard or Voice of the Customer Matrix template.
- *Reduce defects* using the Pareto, Ishikawa, or Fishbone Diagram and Countermeasures Matrix template.
- *Reduce delay* using the Value Stream Map, Flowchart, Value Added Flow Analysis, and Time Tracking templates.
- *Reduce variation* using the control charts and histograms.
- *Reduce measurement error* using the Gage R&R template.
- *Design for breakthrough improvement* using the Failure Modes and Effects Analysis (FMEA), QFD House of Quality, Pugh Concept Selection Matrix, and Design of Experiments.
- *Project management and planning* using the Gantt Chart, Action Plan, and ROI Calculator.

Put Your Whole Improvement Project in One Workbook

Because the QI Macros are an all-in-one toolkit for breakthrough improvement, you can put your entire improvement story in one Excel workbook simply by adding worksheets. Let's say that you've created a control and Pareto chart in one workbook. After you choose "Ishikawa/Fishbone" from the "Fill-in-the-Blanks" templates, just right-click on the sheet name and select "Move" or "Copy Sheet" to move the template into the existing workbook. It's a great way to keep all your information in one place.

Data Transformation

Convert Tables of Data from One Size to Another

Sometimes data has to be reorganized or summarized before it can be graphed. What do you do when your gage or database gives you a single column of data that actually represents several samples (Fig. 8.18)? How do you convert it to work with the XbarR or other chart?

1. Select the single column of data (columns A and B).
2. Click on "Data Transformation–Restacking" to choose various tools, including "Unstack by Label."
3. The macro will reformat your data into three columns and however many rows (columns D through F).

	A	B	C	D	E	F
1	**Drug**	**Diffrate**		Drug 1	Drug 2	Drug 3
2	Drug 1	8		8	10	8
3	Drug 1	4		4	8	6
4	Drug 1	0		0	6	4
5	Drug 1	14		14	4	15
6	Drug 1	10		10	2	12
7	Drug 1	6		6	0	9
8	Drug 2	10				
9	Drug 2	8				
10	Drug 2	6				
11	Drug 2	4				
12	Drug 2	2				
13	Drug 2	0				
14	Drug 3	8				
15	Drug 3	6				
16	Drug 3	4				
17	Drug 3	15				
18	Drug 3	12				
19	Drug 3	9				

Figure 8.18 Single column data.

Summarize Your Data with PivotTables

The QI Macros will draw graphs, but they won't summarize your data because they cannot read your mind. However, you can use Excel's "PivotTable" function to summarize data in almost any conceivable way.

For example, what if you have a series of report codes from a computer system or machine? You need to summarize them before you chart them. Just select the raw data, and go to Excel's menu bar and choose "Data–PivotTable." With a little tinkering, you'll learn how to summarize your data any way you want it.

With the QI Macros, it's easy to create a PivotTable. Just use the mouse to select up to four headings in the sheet (select one heading and then hold down the CTRL key to select up to three more). Then run "Data Transformation–PivotTable Wizard." The wizard will guess how best to organize the data selected into a PivotTable.

Or, using Excel, you can do it the manual way:

1. *Select the labels and data to be summarized* (Fig. 8.19), in this case, denied charges by date, facility, and region. Many computer systems produce one code or measurement each time an event happens. These often need to be summarized to simplify your analysis.
2. *Using the QI Macros, click on up to four column headings, and choose "Data Transformation–PivotTable Wizard"* (Fig. 8.20) to get the PivotTable (Fig. 8.21).

	A	B	C	D	E	F	G	H	I	J	K
1	Region	POST DATE	ENT	ADM DATE	DIS DATE	AS	COS	FC	IN1	PT	DENIED CHARGES
2	North	6/27/03	Hosp1	2/13/03	1/1/00	OL		X	AEH	O	543.07
3	South	12/24/02	Hosp2	7/13/02	1/1/00	OL		X	BCP	E	215.4
4	South	2/25/03	Hosp2	12/6/02	1/1/00			X	CGH	O	157.92
5	South	5/23/03	Hosp3	10/20/02	1/1/00	OL		X	MAH	O	90.73
6	North	7/15/03	Hosp1	5/7/03	1/1/00	AP		X	HEH	O	4103.78
7	North	11/5/02	Hosp4	8/6/01	1/1/00	OL		F	PTB	E	3224.83
8	North	11/20/02	Hosp5	4/15/02	1/1/00	OL		F	PTB	O	3291.76
9	North	11/27/02	Hosp1	5/13/02	1/1/00	OL		F	PTB	O	13845.9
10	North	11/27/02	Hosp4	9/16/02	1/1/00			F	PTB	O	1151

Figure 8.19 PivotTable data.

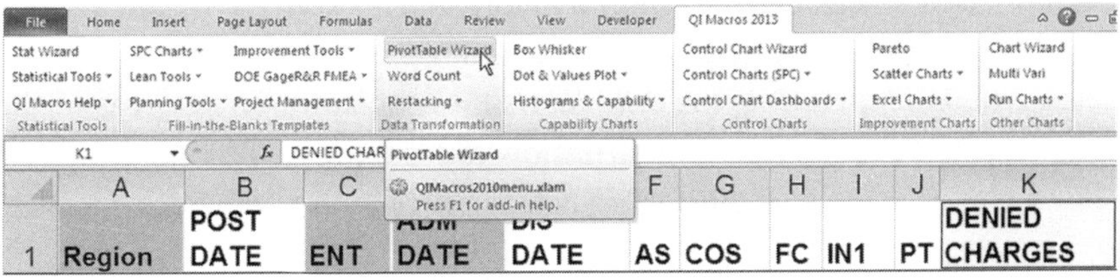

Figure 8.20 QI Macros PivotTable wizard.

	A	B	C	D	E	F	G	H	I
1	Region	(All)							
2									
3	Sum of DENIED CHARGES	ENT							
4	ADM DATE	Hosp1	Hosp2	Hosp3	Hosp4	Hosp5	Hosp6	Hosp7	Grand Total
5	3/28/00			387.48					387.48
6	4/25/00			379.62					379.62
7	3/13/01			6908.98					6908.98
8	7/24/01		311.16						311.16
9	7/26/01					2124.86			2124.86
10	8/6/01				3224.83				3224.83
11	8/20/01		193.65	343.51					537.16
12	10/23/01			230.42					230.42
13	11/16/01			2186.16					2186.16
14	11/19/01			2627.84					2627.84
15	11/26/01			311.2					311.2

Figure 8.21 PivotTable of denied charges.

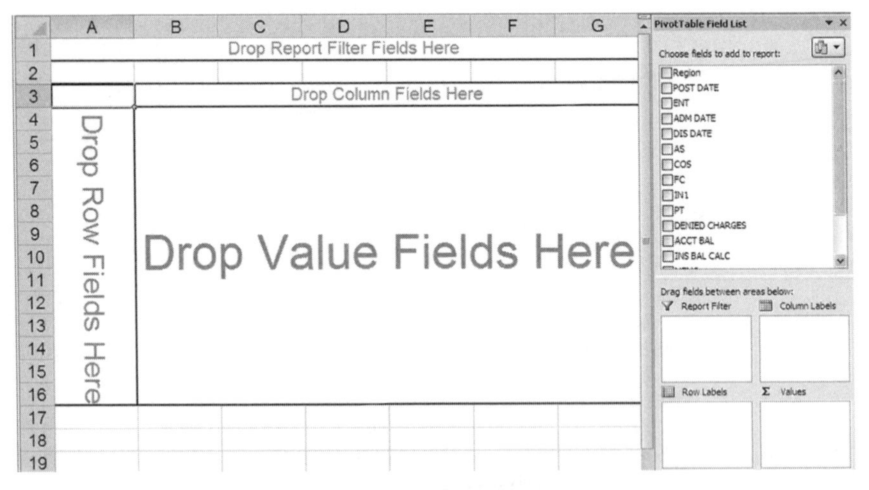

Figure 8.22 Blank PivotTable.

3. In native Excel, choose "Insert–PivotTable" (Excel 2007–2013). Click "Finish" to get a screen like the one in Figure 8.22.
4. *Click, hold, and drag the data labels* into the appropriate areas of the PivotTable to get the summarization you want (Fig. 8.21).
    - *Page fields.* Use this for big categories (e.g., vendor codes, facilities in a company).
    - *Left column.* Use this to summarize by dates or categories.

- *Top row.* Summarize by subcategories.
- *Center.* Drop fields to be counted, summed, or averaged into the center.

5. *To change how the data is summarized*, use the "PivotTable Wizard," or double-click on the top left-hand cell. For online tutorials, Google "Excel PivotTable."
6. *Select labels and totals, and draw charts* using your summarized data.

Using Statistical Tools

Manufacturing companies get into more detailed analysis of data to determine the variation. ANOVA (or analysis of variance) seeks to understand how data is distributed around a mean or average. You can use any of the statistical analysis tools of Excel through the QI Macros. To perform ANOVA in native Excel, you must have Excel's "Data Analysis Toolpak" installed. Go to Excel "Options–Add-Ins–Manage Excel Add-Ins Go" and check "Analysis Toolpak" (Fig. 8.23), and Excel will either turn these tools on or ask you to install them using your Office or Excel CDs. To check if they have been installed, click on "Tools–Data Analysis." If you cannot see "Data Analysis" in the Tools menu, the statistical analysis tools are not installed.

1. Select the data to analyze. This data must be organized in columns.
2. From the QI Macros pull-down menu, select "Statistical Tools" and "Other Analysis Tools" (Excel 2000–2003).
3. Click on the appropriate analysis tool (ANOVA, regression, *f*-test, *t*-test, etc.).

See the sample test data for each tool and test on your computer at c:\qimacros\testdata\anova.xls. (See Chapter 12.)

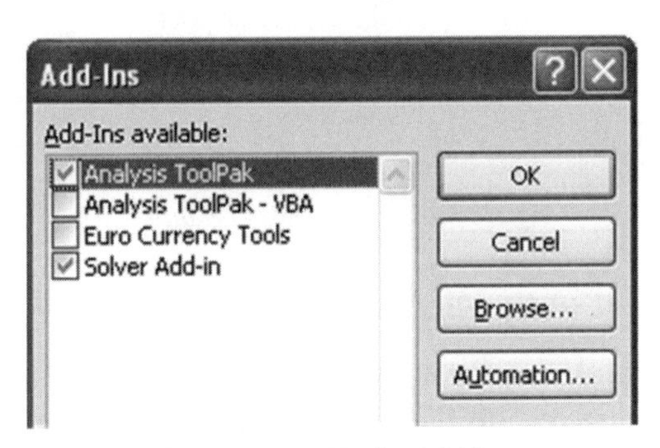

Figure 8.23 Tools–Add-Ins.

Analyzing Customer-Service Data Hidden in Trouble-Reporting Systems

In service industries, much of the information you need to make breakthrough improvements is buried in trouble-reporting systems. Call-center personnel routinely attempt to capture customer complaints, categorize them, and include remarks about customers' dilemmas. Unfortunately, the categories in most information systems are predefined, inflexible, and rarely speak to the true nature of a customer's complaint. And often the customer, who has waited in a call queue for several minutes, has had time to think up several questions that he or she needs answered, not just one (Fig. 8.24).

In these situations, the information needed to analyze these customer interactions is in the freeform remarks, not in convenient categories. The information captured in the remarks invariably will be more accurate than that in the predefined categories. How do you analyze this wild potpourri of short phrases and abbreviations? The answer lies in Microsoft Excel.

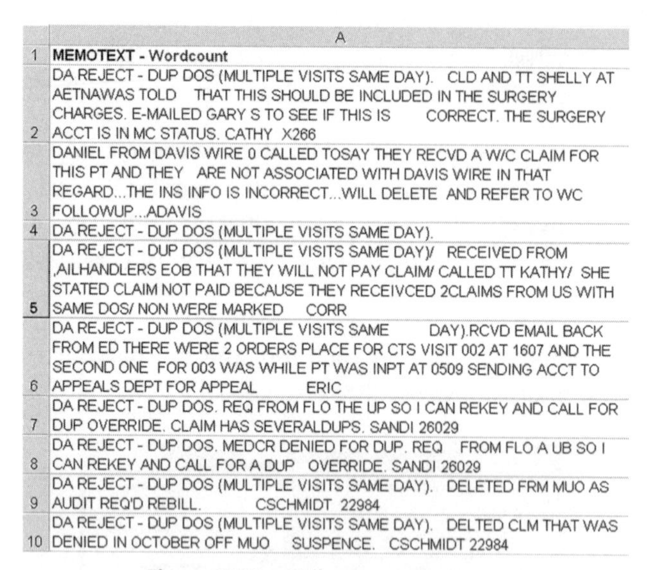

	A
1	**MEMOTEXT - Wordcount**
2	DA REJECT - DUP DOS (MULTIPLE VISITS SAME DAY). CLD AND TT SHELLY AT AETNAWAS TOLD THAT THIS SHOULD BE INCLUDED IN THE SURGERY CHARGES. E-MAILED GARY S TO SEE IF THIS IS CORRECT. THE SURGERY ACCT IS IN MC STATUS. CATHY X266
3	DANIEL FROM DAVIS WIRE 0 CALLED TOSAY THEY RECVD A W/C CLAIM FOR THIS PT AND THEY ARE NOT ASSOCIATED WITH DAVIS WIRE IN THAT REGARD...THE INS INFO IS INCORRECT...WILL DELETE AND REFER TO WC FOLLOWUP...ADAVIS
4	DA REJECT - DUP DOS (MULTIPLE VISITS SAME DAY).
5	DA REJECT - DUP DOS (MULTIPLE VISITS SAME DAY)/ RECEIVED FROM ,AILHANDLERS EOB THAT THEY WILL NOT PAY CLAIM/ CALLED TT KATHY/ SHE STATED CLAIM NOT PAID BECAUSE THEY RECEIVCED 2CLAIMS FROM US WITH SAME DOS/ NON WERE MARKED CORR
6	DA REJECT - DUP DOS (MULTIPLE VISITS SAME DAY).RCVD EMAIL BACK FROM ED THERE WERE 2 ORDERS PLACE FOR CTS VISIT 002 AT 1607 AND THE SECOND ONE FOR 003 WAS WHILE PT WAS INPT AT 0509 SENDING ACCT TO APPEALS DEPT FOR APPEAL ERIC
7	DA REJECT - DUP DOS. REQ FROM FLO THE UP SO I CAN REKEY AND CALL FOR DUP OVERRIDE. CLAIM HAS SEVERALDUPS. SANDI 26029
8	DA REJECT - DUP DOS. MEDCR DENIED FOR DUP. REQ FROM FLO A UB SO I CAN REKEY AND CALL FOR A DUP OVERRIDE. SANDI 26029
9	DA REJECT - DUP DOS (MULTIPLE VISITS SAME DAY). DELETED FRM MUO AS AUDIT REQ'D REBILL. CSCHMIDT 22984
10	DA REJECT - DUP DOS (MULTIPLE VISITS SAME DAY). DELTED CLM THAT WAS DENIED IN OCTOBER OFF MUO SUSPENCE. CSCHMIDT 22984

Figure 8.24 Call center comments.

Importing Text with Microsoft Excel

To analyze text with Excel, you must first import the text into Excel. To do this, you will need to export the customer account and remarks information from the trouble-reporting system into your PC or local area network.

To simplify deeper analysis, it will be useful to have something about the customer's account included with the remark. In a phone company, for example, having the customer's phone number will enable further analysis by digging into the customer's records.

Then go to Excel and choose "File–Open," select "Files of Type: All Files," and open the text file. Excel's Import Wizard will then guide you through importing the data. Text data can either be *delimited*, which means it contains tab, comma, or other characters that separate fields, or *fixed width*, which means that the data is of a consistent length.

The maximum number of characters Excel will store in a cell is 255, so longer text fields should be edited to fit. More than one cell can be used to store an entire remark or comment. Excel will allow 65,000 or more rows to be imported per Excel worksheet.

Analyzing Text with Excel

Call centers often collect vital information necessary for improvement. Searching the call-center comments in an imported text file couldn't be easier. For sample data, look in the QI Macros "Test Data" for pivottable.xls, and click on the "Word Count" sheet.

Native Excel has a function called "COUNTIF" that tallies cells if they match certain criteria. The formula for the "COUNTIF" function is

=COUNTIF(CellRange, "criteria")

The "CellRange" specifies the range of cells to be counted. If there is only a single column of imported text, this might be "$A$2:$A$10" (Fig. 8.25).

B2 =COUNTIF(A2:A10,"=*"&B1&"*")

	A	B
1	**MEMOTEXT** - Wordcount	DUP DOS
2	DA REJECT - DUP DOS (MULTIPLE VISITS SAME DAY). CLD AND TT SHELLY AT AETNAWAS TOLD THAT THIS SHOULD BE INCLUDED IN THE SURGERY CHARGES. E-MAILED GARY S TO SEE IF THIS IS CORRECT. THE SURGERY ACCT IS IN MC STATUS. CATHY X266	8
3	DANIEL FROM DAVIS WIRE 0 CALLED TOSAY THEY RECVD A W/C CLAIM FOR THIS PT AND THEY ARE NOT ASSOCIATED WITH DAVIS WIRE IN THAT REGARD...THE INS INFO IS INCORRECT...WILL DELETE AND REFER TO WC FOLLOWUP...ADAVIS	

Figure 8.25 Word count of comments.

Once you've specified the range, the real trick is to create criteria consisting of keywords and phrases that match the cells. To do this, you'll need to use Excel's "wildcard" character, the asterisk (*). To match a cell that contains a keyword, the criteria portion of the "COUNTIF" statement will need to look for any leading stream of characters (*), the keyword, and any trailing stream of characters (*). The simple way of expressing this in the "COUNTIF" statement would be

=COUNTIF(CellRange,"=*_keyword_*")

To make this easy to change, you might consider putting the keyword in one cell by itself and including it into the *formula*. The formula would be

=COUNTIF($A$1:$A$2154,"=*"&B1&"*")

This would take the keyword from the cell above it, making it easier to change and test various keywords. Getting the keyword right can make the resulting data more accurate.

QI Macros Word Count Tool

Word Count parses the words out of sentences and paragraphs and uses PivotTables to count the occurrences of individual words and sort them in descending order (Figure 8.26). To count the words in your selection:

1. Select the cells you want to analyze (in this case, "Word Count" sheet in pivottable.xls).
2. Click on QI Macros "Data Transformation—Word Count."

	A	B	C	D	E
1	Count of Word			Count of Two-Word Phrases	
2	Word	Total		Two-Word Phrases	Total
3	DUP	11		REJECT DUP	8
4	DOS	9		DUP DOS	8
5	REJECT	8		DA REJECT	8
6	DA	8		MULTIPLE VISITS	6
7	SAME	7		DOS MULTIPLE	6
8	FOR	7		VISITS SAME	6
9	MULTIPLE	6		SAME DAY	5
10	FROM	6		FOR DUP	3

Figure 8.26 QI Macros word count of key words and phrases.

Word Count then will parse each word out of each cell and summarize and order them using Excel's PivotTable function.

From these comments, the most likely cause of rejected claims was determined to be overlapping dates of service (DOS).

Graphing the Data

Once you've mined all the data out of the comments, you can then use Pareto charts to examine the frequency of certain types of customer complaints. Additional digging into specific customer records may be required to determine the root cause of why these calls are being generated and how to mistake-proof the process to prevent them.

Power Tools for Breakthrough Improvement

As you can see from these examples, Excel and the QI Macros are power tools to simplify breakthrough improvement. By putting your data into Excel, summarizing it with PivotTables, and graphing it with the QI Macros, you can automate and accelerate your journey toward breakthrough improvement.

1. The QI Macros give you the power to select data and immediately draw all the key charts and diagrams necessary for breakthrough improvement—Pareto charts and fishbone diagrams for problem solving as well as histograms and control charts for monitoring the process.
2. The QI Macros templates give you fill-in-the-blanks simplicity for control charts, Pareto charts, fishbone diagrams, flowcharts, and value-stream mapping.
3. The QI Macros ANOVA and analysis tools give you simplified access to Excel's statistical tools and much more.
4. Control Chart Dashboards simplify monthly reporting.
5. Data-transformation tools help to reorganize or PivotTable the data.

Start using Excel and the QI Macros to organize, analyze, and graph your data to illuminate the opportunities for improvement.

Troubleshooting Problems

Users have three types of questions when using the QI Macros.

1. *Statistical process control questions.* What chart should I use? If you use the Chart Wizard in the QI Macros, the software will choose your chart for you. Otherwise, most of these questions are answered on my website www.qimacros.com/support/spc-faqs/.
2. *Excel questions.* How do I enter my data? Why don't I get the right number of decimal places? Most of these are answered at www.qimacros.com/support/excelfaq.
3. *QI Macros/Excel/Windows support issues.* Most of these are answered at www.qimacros.com/support.

Here are some common issues:

- *How do I set up my data?* See examples in the QI Macros Test Data.
- *Decimal points* (e.g., 12.02). Excel stores most numbers as General format. To get greater precision, simply select your data and go to "Format Cells–Number" to specify the number of decimals. Then run your chart.
- *Headers shown as data.* Are your headers numerical? If so, you need to put an apostrophe (') in front of each heading.
- *No data (one cell), too much data (entire columns/rows), or the wrong data selected.* Are just the essential data cells highlighted?
- *Data in text format.* Are your numbers left aligned? To convert to numbers, simply put the number 1 in a blank cell, select "Edit–Copy," then select your data, and choose "Paste–Special–Multiply."
- *Hidden rows or columns of data.* Users sometimes hide a column or row of data in Excel (e.g., columns show A, B, and then F). If you select "A–F," you get all the hidden data too!
- *Data in the wrong order.* Some of these macros require two or more columns of data. The *p* chart expects (1) a heading, (2) the number of defects, and (3) the sample size. If columns 2 and 3 are reversed, it won't work properly.
- *To uninstall the macros,* download www.qimacros.com/uninstall.

Technical Support

If you're still having problems, check out www.qimacros.com/support or e-mail your Excel file and problems to support@qimacros.com. Include the version number and service pack of Excel and Windows or MacOS.

CHAPTER 9

Breakthrough Improvement Using Microsoft Excel

To be effective, analytics tools must be simple, powerful, and usable by decision makers, not just IT experts.

—CHARLES BREWER, CEO NAMAX DI, LTD.

Recently I've noticed that one of the biggest challenges improvement professionals face is figuring out how to develop improvement stories from volumes of data. In most cases, the data looks like Figure 9.1—rows and rows of detail about mistakes, errors, or scrap. It shows dates, machines, reason for scrap, and total weight and scrap weight in pounds.

This kind of data reminds me of the old joke about the father who finds his daughter digging through a pile of horse manure. When he asks her what she's doing, she replies, "There has to be a pony in here somewhere!"

	A	B	C	D	E
1	Date	Machine	Reason for Scrap	Weight	Scrap Weight
2	2007-01-02	M10	Dents	5,832	266
3	2007-01-02	M10	Dirty Metal	7,676	72
4	2007-01-02	M10	Wrong Incoming Jevils	7,358	399
5	2007-01-02	M4	Mechanical Problems	10,733	2,018
6	2007-01-02	M5	Tracking	8,039	681
7	2007-01-02	M6	Arbor Break	8,448	218
8	2007-01-02	M6	Arbor Break	8,584	409
9	2007-01-02	M8	Product Change	6,903	2,613
10	2007-01-02	M9	Sheet Walked Off	4,718	345

Figure 9.1 Scrap metal data.

Is there an improvement pony in your pile of data? With data like this, you will want to do some data mining with Excel to find the breakthrough improvement projects hidden in your data—the invisible low-hanging fruit.

Data Mining with Excel

Using Excel's "PivotTable" function, it's easy to analyze this data to find improvement stories. Here's the process:

1. Click on any cell in the data. (Excel will automatically select all rows and columns in your data.)
2. Click on "Insert–PivotTable" (Fig. 9.2).

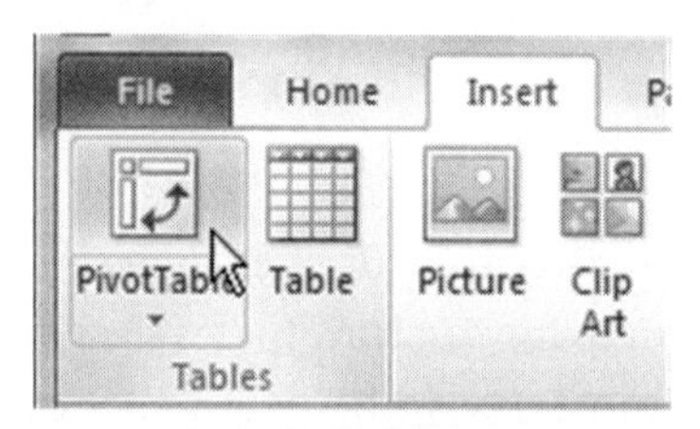

Figure 9.2 Insert PivotTable.

Excel will prompt with default choices for creating a PivotTable (Fig. 9.3).

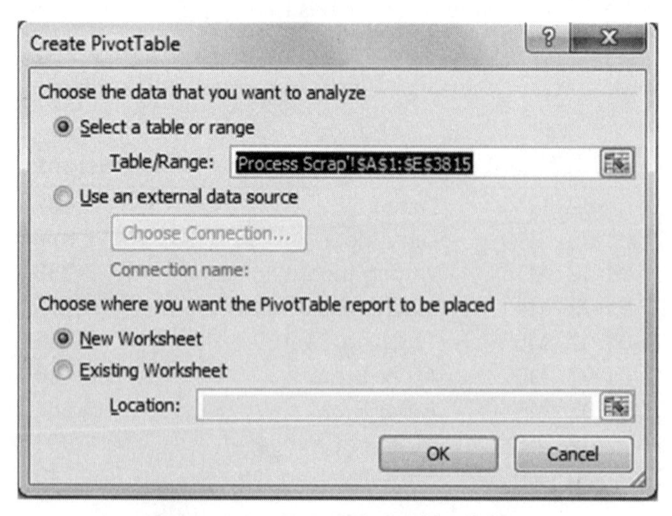

Figure 9.3 Create PivotTable dialog.

Click "OK" and "Finish" to reveal the PivotTable layout (Fig. 9.4).

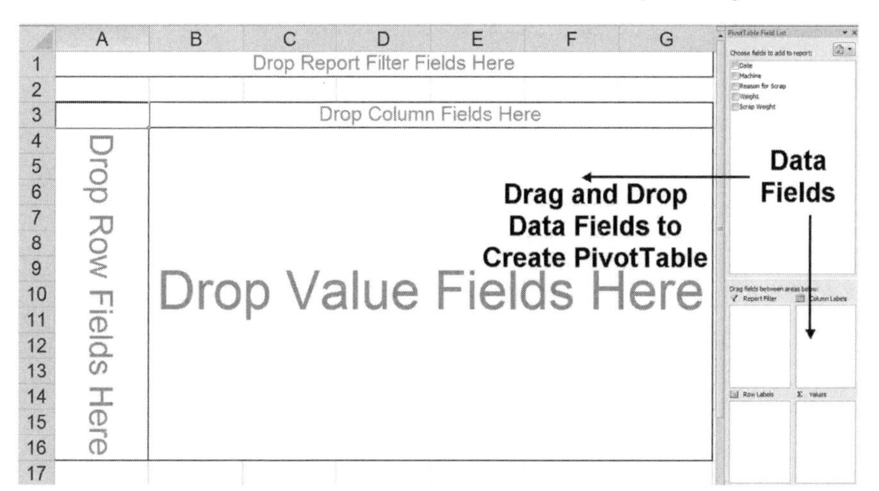

Figure 9.4 PivotTable layout.

3. Now simply use your mouse to drag and drop items onto the PivotTable fields.

TIP Use only one data field per area; otherwise, it will be hard to create charts from the resulting PivotTable.

---

a. *Want scrap by day?* Drag the date into "Drop Row Fields Here," and drag scrap weight into "Drop Value Fields Items Here" (Fig. 9.5).

	A	B
1		
2		
3	**Row Labels**	**Sum of Scrap Weight**
4	01/02/07	7219
5	01/03/07	468
6	01/04/07	3568
7	01/05/07	4403
8	01/06/07	8370
9	01/07/07	77
10	01/08/07	910

Figure 9.5 Scrap weight by date.

b. *Want scrap by machine?* Drag machine into "Drop Row Fields Here," and drag scrap weight into "Drop Value Fields Here." Then sort the scrap in descending order by clicking on the "Total" column and "Data Sort–Descending" (Fig. 9.6).

	A	B
1		
2		
3	Sum of Scrap Weight	
4	Machine	Total
5	M1	4155939
6	M8	525740
7	M6	444110
8	M10	328076
9	M9	256971
10	M3	189165
11	M4	143961
12	M2	49113
13	M5	33581
14	M11	4554
15	M7	1352
16	Grand Total	6132562

Figure 9.6 Scrap by machine.

c. *Want scrap by reason for scrap and machine?* Drag reason for scrap into "Drop Row Fields Here," drag machine into "Drop Column Field Here," and drag scrap weight into "Drop Value Fields Here." Then sort into descending order (Fig. 9.7).

If you want to see the data that makes up any cell in the PivotTable, just double-click on the cell. If you click on cell B3 ("Surface Cracks" and "M1"), you get Figure 9.8.

	A	B	C	D	E	F	G	H	I	J	K	L	M
1	Sum of Scrap Weight	Machine											
2	Reason for Scrap	M1	M10	M11	M2	M3	M4	M5	M6	M7	M8	M9	Grand Total
3	Surface Cracks	1098637			9186	6448	2208						1116479
4	Warm Up	799159			1243	2423	5592						808417
5	Transistion	720940											720940
6	Edge Cracks	366382	1433		7792	41003	25344		15064				457018
7	Bad Shape	374049							4725			401	379175
8	Electrical Problem	116562	4916			1751	12582	3625	390		85235	3940	229001
9	Oscillation	81250	29820		1610	4508	12987	1430	3023		19347	16293	170268
10	Sheet Walked Off	6275	16364		6578	11057	32436		4070		72598	20657	170035

Figure 9.7 Scrap by reason and machine.

	A	B	C	D	E
1	Date	Machine	Reason for Scrap	Weight	Scrap Weight
2	2007-05-17	M1	Surface Cracks	8465	8465
3	2007-05-17	M1	Surface Cracks	9235	9235
4	2007-05-17	M1	Surface Cracks	2080	2080
5	2007-05-17	M1	Surface Cracks	9835	9835
6	2007-05-15	M1	Surface Cracks	8840	8840
7	2007-05-15	M1	Surface Cracks	9895	9895
8	2007-05-14	M1	Surface Cracks	10695	10695
9	2007-05-14	M1	Surface Cracks	9455	9455
10	2007-05-13	M1	Surface Cracks	5730	5730

Figure 9.8 Scrap by reason for Machine 1 (M1).

This ability to summarize and then bring up the data behind each cell is the essence of data mining in Excel.

QI Macros PivotTable Wizard: The Shortcut to Breakthrough Improvement

If you are like many Excel users, you struggle with creating PivotTables in Excel. However, PivotTables are a valuable tool that every breakthrough improvement expert should know how to use. The QI Macros makes creating PivotTables easy. Here is how:

1. Perform some basic data cleanup on your data (see Fig. 9.1).
   a. Make sure that each column has a distinct heading.
   b. Check for misspellings or inconsistencies in your data. Use Excel's "Find and Replace" function to make these consistent.
2. Next click on one to four column headings to select them. These columns will be included in your PivotTable. Using the data in Figure 9.1, we could select "Machine," "Reason for Scrap," and "Scrap Weight."
3. Next, click on the QI Macros menu, and choose "PivotTable Wizard" (Fig. 9.9).
4. Based on the frequency of values in each column, the QI Macros PivotTable Wizard figures out where to place each slice of data and automatically creates a PivotTable for you. You may then need to move fields around a little to get exactly what you want (Fig. 9.10).

Machine	Reason for Scrap	Weight	Scrap Weight
M10	Dents	5,832	266
M10	Dirty Metal	7,676	72

Figure 9.9 QI Macros PivotTable Wizard.

Sum of Scrap Weight	Machine											
Reason for Scrap	M1	M10	M11	M2	M3	M4	M5	M6	M7	M8	M9	Grand Total
Anneal Stain		1289										1289
Arbor Break		5326						16351			6671	28348
Bad Coating		495									15112	15607
Bad Shape	374049							4725			401	379175
Bad Trim		3337		6888	18994			9370			2357	40946
Bridle Kicked Out								6754		336		7090
Buckles		1972			812			7653			8031	18468
Chatter								1345				1345
Chemistries Off Grade	30860											30860
Chopper Plugged				2448	8179	16391						27018
Coating Pattern		688								7020		7708

Figure 9.10 PivotTable Wizard results.

Breakthrough Improvement Projects

I usually create control and Pareto charts as I'm mining the data from PivotTables. From my perspective, most breakthrough improvement stories consist of using four key tools in the right order.

1. *PivotTable* of defect, mistake, error, scrap, or cost data
2. *Control chart* of performance over time
3. *Pareto chart* (two or more levels of detail)
4. *Ishikawa or fishbone diagram* of root-cause analysis

So here's how you could analyze this scrap data using these tools. We've already used PivotTables to summarize the data, so we use data from Figure 9.5 to create an XmR control chart (Fig. 9.11).

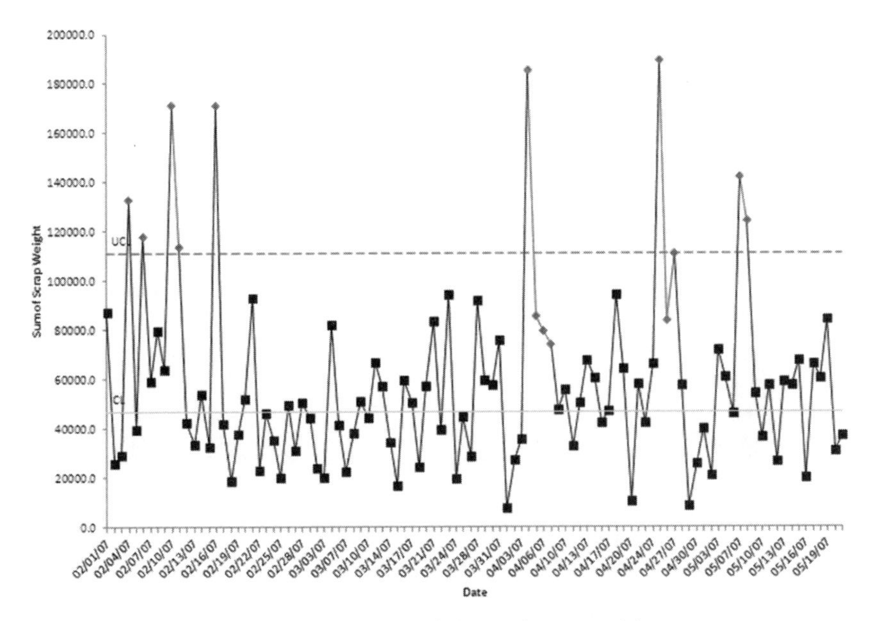

Figure 9.11 Control chart of scrap weight.

Then we use the data in Figure 9.6 to create a Pareto chart (Fig. 9.12).
Because Machine 1 causes over 67 percent of the scrap (the 4-50 Rule), we double-click on on the "Grand Total" for Machine 1 in the PivotTable

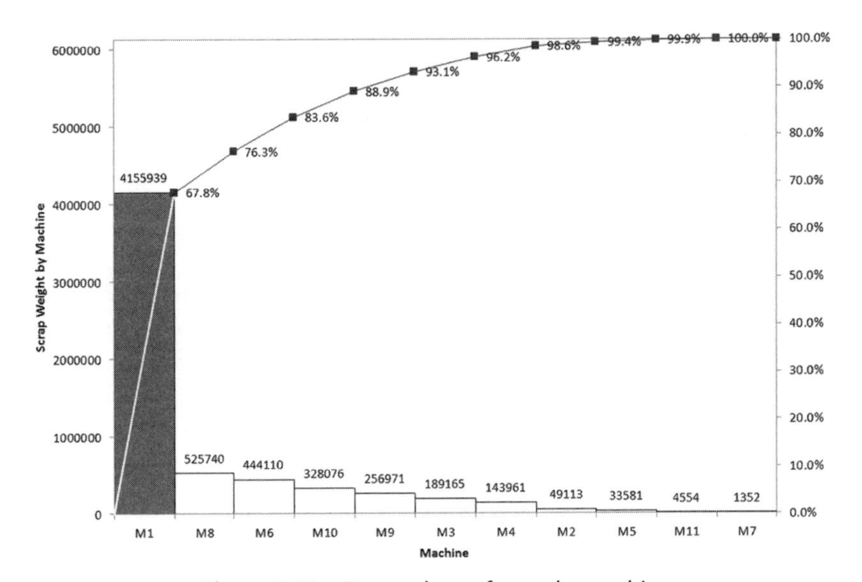

Figure 9.12 Pareto chart of scrap by machine.

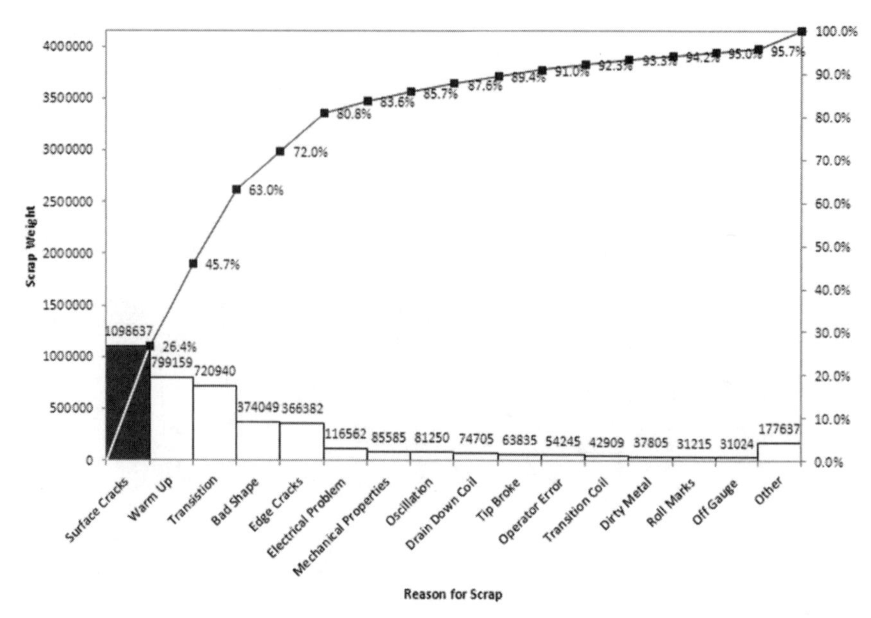

Figure 9.13 Pareto chart of reason for scrap on machine 1.

and then create a PivotTable of "Reasons for Scrap on Machine 1" and a Pareto chart of the reasons (Fig. 9.13).

Surface cracks are the biggest reason for scrap on Machine 1, so we can create an Ishikawa (fishbone) diagram as a starting point for root-cause analysis (Fig. 9.14).

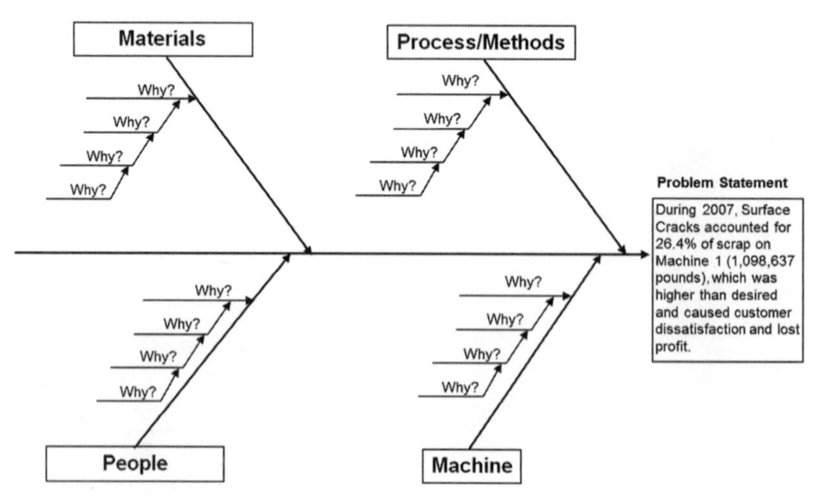

Figure 9.14 Ishikawa diagram for root-cause analysis.

Using this analysis, it's a lot easier to figure out who should be on this breakthrough improvement team (employees who work on Machine 1 and understand surface cracks).

If you really wanted to make dramatic and rapid improvement, you could add root-cause teams for all of the top five bars of the Pareto chart: "Warm Up," "Transition," "Bad Shape," and "Edge Cracks." Having multiple teams focused on solving the top problems ensures breakthrough improvement. This is the essence of data mining and data analysis.

Small Data Analytics for Breakthrough Improvement

There's a lot of buzz in industry about *data analytics*—mining huge data sets to discover invisible patterns of customer behavior that can be leveraged to maximize sales. PivotTables are the key to small data analytics. So let's explore how to use PivotTables to analyze another data set.

Over the years, I have used Excel PivotTables on projects that:

- Saved $20 million in postage expense
- Saved $16 million in adjustment costs
- Reduced order errors in a wireless company from 17 to 3 percent in just four months, saving $250,000 a month in rework
- Reduced denied insurance claims in a health care system that saved $5 million a year with simple process changes that could be implemented immediately

Using Excel PivotTables for data mining is the key to finding multimillion-dollar improvement projects.

What Small Data Looks Like

Almost every business I've ever consulted with has data about defects, mistakes, and errors in an Excel spreadsheet. The data is a line-by-line, date-by-date account of the origin and type of defect (Fig. 9.15).

For lost-time analysis (LTA), the data might look like Figure 9.16.

	A	B	C
1	Date	Line	Defect - Mistake - Error Type
2	1/2/2010	Line 2	Bent/Damaged flaps
3	1/2/2010	Line 2	Carton will not open
4	1/2/2010	Line 3	Folded flaps
5	1/2/2010	Line 3	Off color
6	1/2/2010	Line 1	Bent/Damaged flaps

Figure 9.15 Manufacturing defects.

	A	B	C
1	Date	Lost Time (minutes)	Cause
2	01/02/12	45	Generator Failure
3	01/03/12	15	Conveyor Failure
4	01/04/12	10	Training
5	01/05/12	5	Safety

Figure 9.16 Lost time data.

Turning Data into Improvement Projects

To turn this kind of data into something that can be control charted and Pareto charted requires the use of Excel PivotTables. PivotTables can

- Count the occurrences of a keyword, phrase, or number
- Sum or average numbers
- Turn raw data into knowledge
- Help drill down into mounds of data

Using the data in Figure 9.15, simply select the three column headings, and choose the QI Macros "PivotTable Wizard" to get a PivotTable (Fig. 9.17).

Now you can select dates in column A and total defects in column E to draw a control chart of defects by day (Fig. 9.18).

In this case, we could remove the dates from the PivotTable and put in the type of defect (Fig. 9.19).

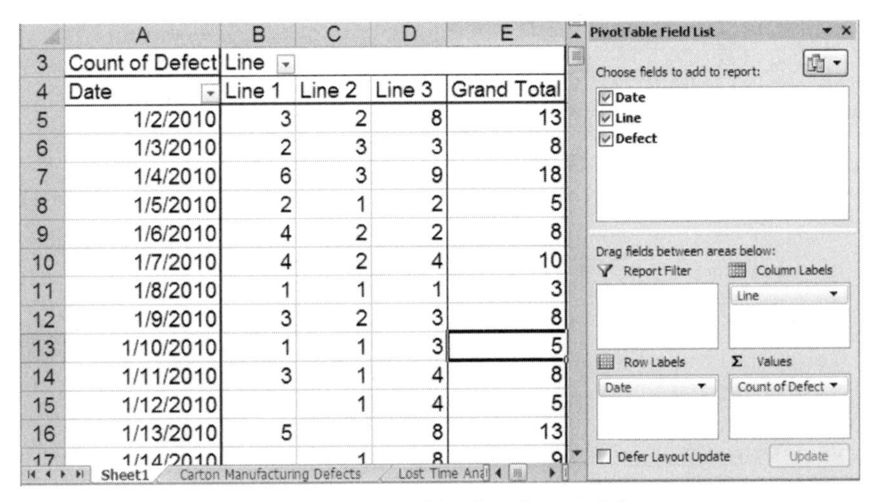

Count of Defect	Line			
Date	Line 1	Line 2	Line 3	Grand Total
1/2/2010	3	2	8	13
1/3/2010	2	3	3	8
1/4/2010	6	3	9	18
1/5/2010	2	1	2	5
1/6/2010	4	2	2	8
1/7/2010	4	2	4	10
1/8/2010	1	1	1	3
1/9/2010	3	2	3	8
1/10/2010	1	1	3	5
1/11/2010	3	1	4	8
1/12/2010		1	4	5
1/13/2010	5		8	13
1/14/2010		1	8	9

Figure 9.17 PivotTable of packaging defects.

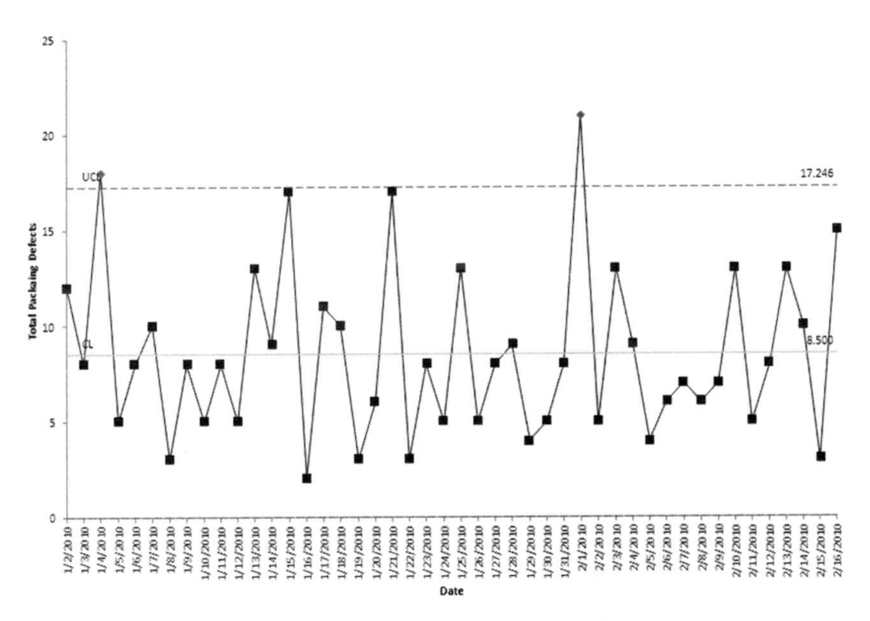

Figure 9.18 C chart of total packaging defects.

	A	B	C	D	E
1	Count of Defect	Line			
2	Defect	Line 1	Line 2	Line 3	Grand Total
3	Bent/Damaged flaps	37	23	24	84
4	Carton will not open	29	18	29	76
5	Damaged Pallet	3			3
6	Fisheye	9			9
7	Folded flaps	16	6	82	104
8	Ink smears/streaks		5	19	24
9	Mislabeled			3	3
10	Missing color			8	8
11	Off color	14	5	12	31
12	Oil spots			14	14
13	Poor ink adhesion	7	8	18	33
14	Undercount		2		2
15	Grand Total	115	67	209	391
16					

PivotTable Field List
Choose fields to add to report:
Date
Line
Defect
Drag fields between areas below:
Report Filter
Column Labels
Line
Row Labels
Defect
Values
Count of Defect

Figure 9.19 Changing PivotTable.

Select B2:D2 and B15:D15 to get a Pareto chart of total defects by line (Fig. 9.20). Line 3 clearly has the most defects.

Then select A2:E14 to get Pareto charts of types defects (Figs. 9.21 through 9.24). On line 3, "Folded Flaps" accounts for 39.2 percent of line 3

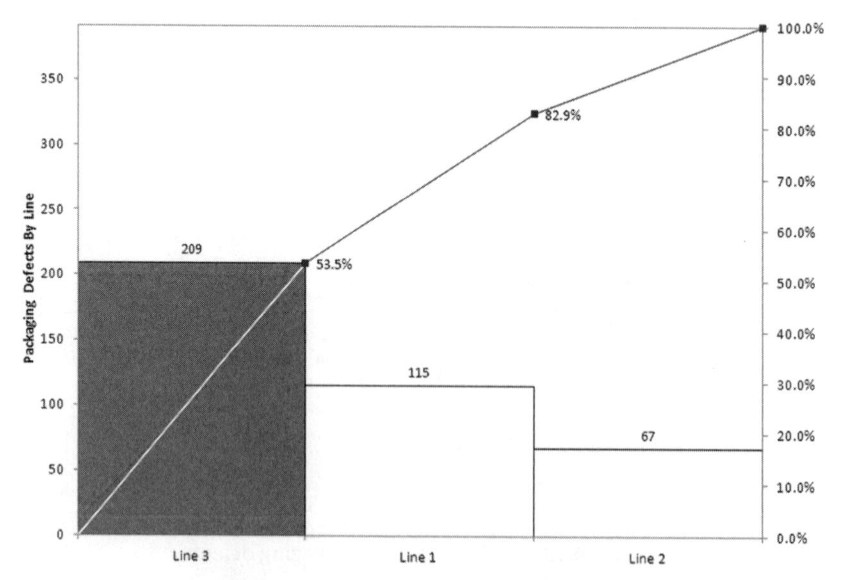

Figure 9.20 Pareto chart of total packaging defects.

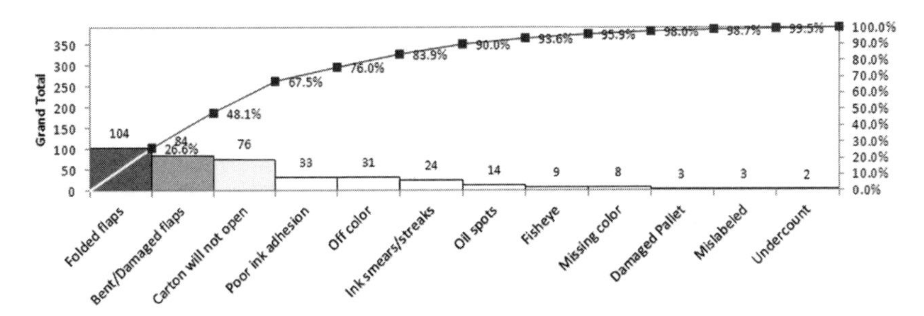

Figure 9.21 Pareto chart of total packaging defects.

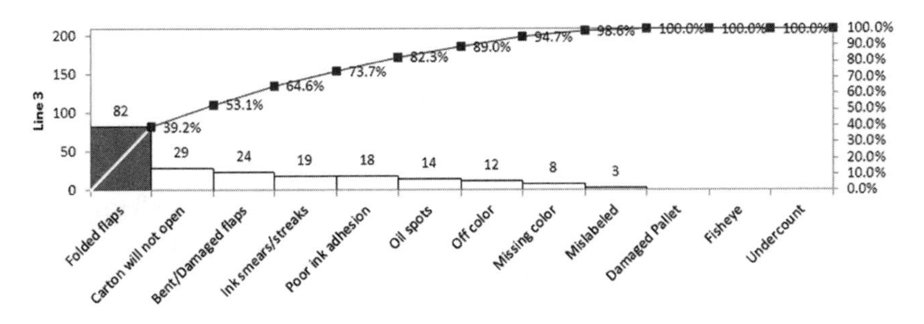

Figure 9.22 Pareto chart of line 3 defects.

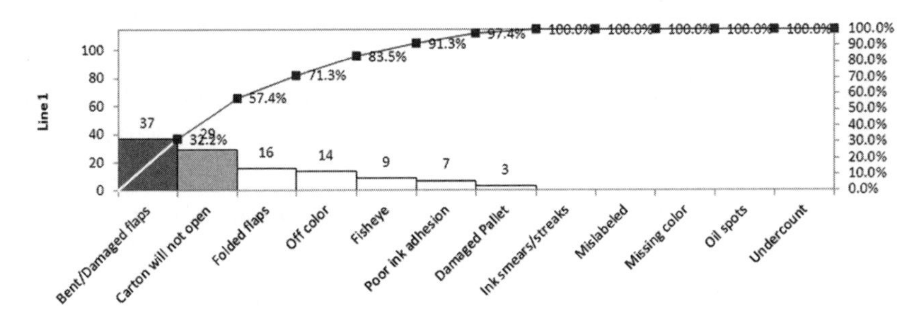

Figure 9.23 Pareto chart of line 1 defects.

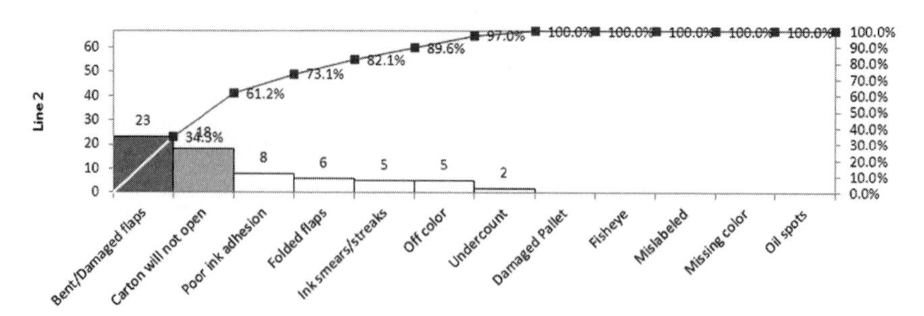

Figure 9.24 Pareto chart of line 2 defects.

defects and is the largest overall problem, followed by "Bent/Damaged Flaps" and "Carton Will Not Open," which are the main problems on lines 1 and 2.

To make breakthrough improvements, the line 3 team can tackle "Folded Flaps," line 1 can tackle "Bent/Damaged Flaps," and line 2 could tackle "Carton Will Not Open." Solutions from each line then can be *replicated* on the other lines.

To learn more about how to use PivotTables for breakthrough improvement, watch the videos at www.qimacros.com/breakthrough-improvement-excel.

Lost Time Analysis

The Lost Time data is very similar. We could use it for computer downtime, power outages, or machine outages. Using the data in Figure 9.17, select the three headings and use the QI Macros PivotTable Wizard to summarize the data (Fig. 9.25).

Using the dates (A3:A27) and Grand Total (H3:H27), you can create a control chart of lost time (Fig. 9.26). In this case, use the XmR chart because time is *measured* data. Average lost time is 20 minutes per failure, but could be as high as 76 minutes. The range chart (below the X chart) shows variation from failure to failure.

Using the column headings (B2:G2) and Grand Total (B28:G28), you can create a Pareto chart (Fig. 9.27) of the main causes of lost time. Generator failure is the number one cause of lost time (346 minutes is 68.4 percent of the total lost time).

Sum of Lost Time (minutes)	Cause						
Date	Conveyor Failure	Flooding	Generator Failure	Pump Failure	Safety	Training	Grand Total
01/02/12			45				45
01/03/12	15						15
01/04/12						10	10
01/05/12					5		5
01/06/12		9					9
01/07/12				9			9
01/08/12			40				40
01/09/12					6		6
01/10/12	15						15
01/11/12			35				35
01/12/12					5		5
01/13/12			30				30
01/14/12		11					11
01/15/12			35				35
01/16/12						11	11
01/17/12			41				41
01/18/12	14						14
01/19/12			45				45
01/20/12					5		5
01/21/12			35				35
01/22/12						13	13
01/23/12	15						15
01/24/12			40				40
01/25/12					5		5
01/26/12	12						12
Grand Total	71	20	346	9	26	34	506

Figure 9.25 Lost Time PivotTable.

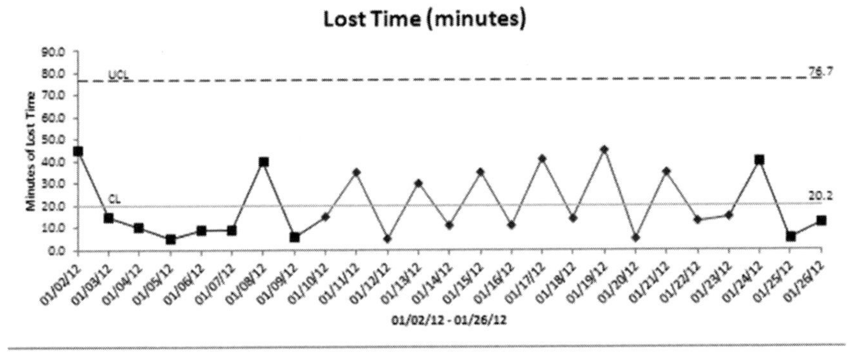

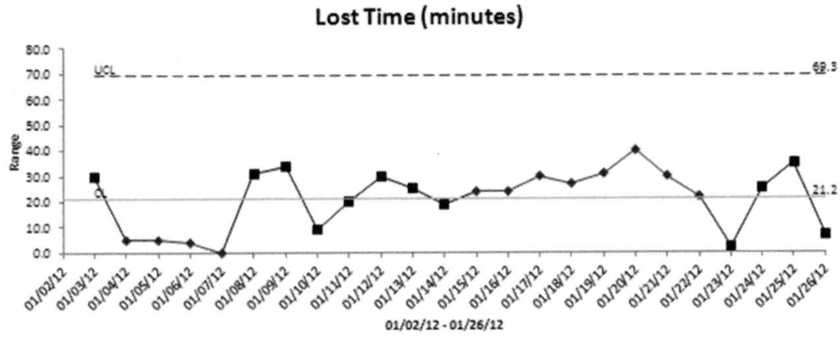

Figure 9.26 Lost Time XmR chart.

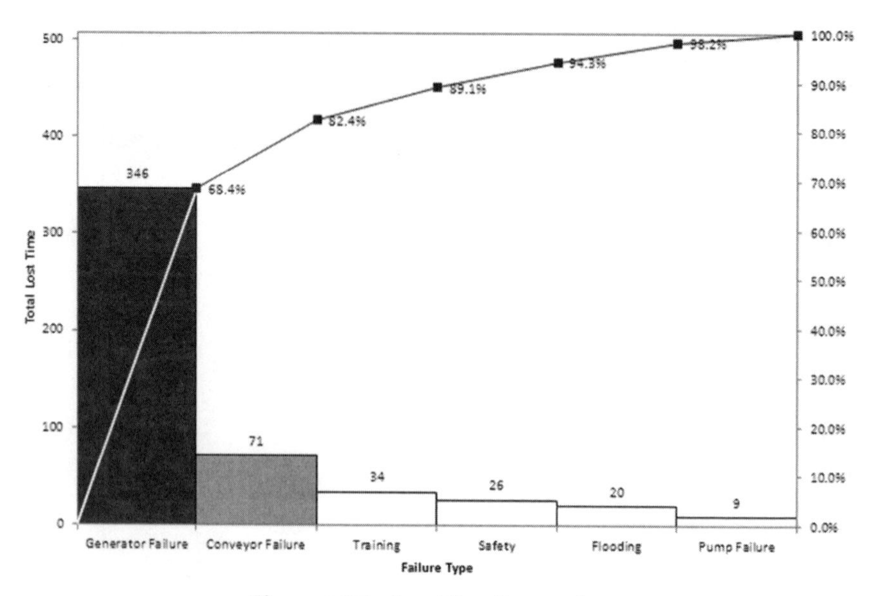

Figure 9.27 Lost Time Pareto chart.

Since there is no additional information about generator failures, the next step is to create an Ishikawa or fishbone diagram (Figure 9.28). Now you know that you need people on the team that are experts at analyzing

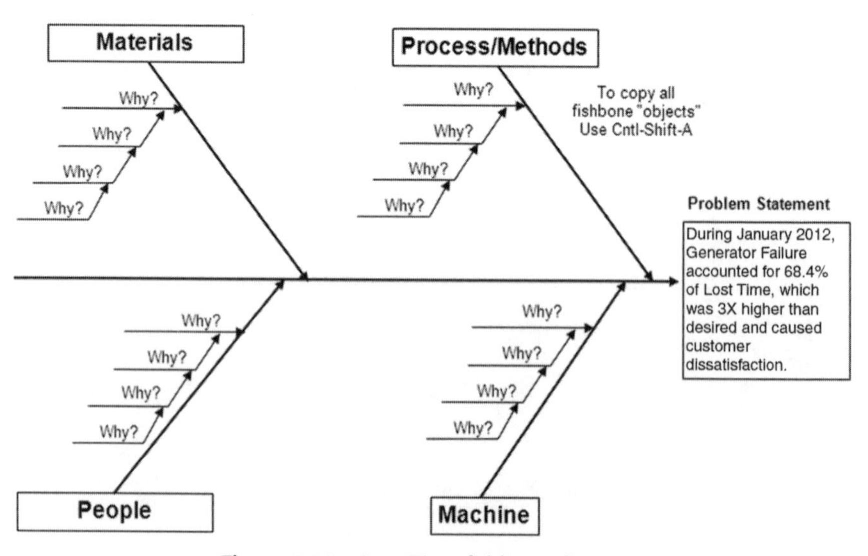

Figure 9.28 Lost Time fishbone diagram.

why the generator fails and how to prevent it (e.g., preventive maintenance, failure analysis and replacement).

Using PivotTables to summarize this data and control charts and Pareto charts to analyze this type of data will quickly narrow your focus to the key problems to solve and the right people to have on the team.

Get the Idea?

There's a wealth of information hiding in these dense files of words and numbers. Start using Excel's "PivotTable" function to slice and dice your data (no matter how large). From there, you can draw control charts and Pareto charts using the QI Macros. And then drill down into any PivotTable cell to find the biggest "pain" and determine how to fix it. It's easy to find the invisible low-hanging fruit—the 4 percent of the business that causes 50 percent of the problems and start making breakthrough improvements.

The data is out there—start digging!

CHAPTER 10

Sustaining Breakthrough Improvement

It's not enough to create breakthrough improvements; you also have to *sustain* the improvement forever. Unfortunately, this is where most improvements fail; teams don't put a system in place to monitor and maintain the new level of performance.

Once you have achieved success, you don't want to backslide. The best way to ensure continued performance is to monitor and track the key measures you identified in your breakthrough improvement analysis. There are two main types of key process indicators (KPIs): quality and process indicators.

Quality and Process Indicators

Process indicators measure performance *inside* the process. They help find and fix problems before the customer is affected. Put them at critical hand-offs and decision points—especially ones that require error detection and correction. Count the number of times through a rework loop or the number of pieces that have to be reworked (e.g., orders, bills, purchases, parts, etc.).

Quality indicators are measured *after* delivery. They track customer satisfaction with timeliness, accuracy, and value. Quality indicators could include complaints, returns, warranty repairs, and so on.

Common and Special Causes

Processes are never perfect. So how can you tell if a process is stable and predictable? Common and special causes of variation make the process perform differently in different situations. Getting from your home to school or work takes varying amounts of time because of traffic or transportation delays. These are *common causes of variation*; they exist every day. A blizzard, a traffic accident, a chemical spill, or some other freak occurrence would be a *special cause of variation*.

It's easy to identify common and special causes using control charts. You don't have to know any of the rules for evaluating control charts because the QI Macros control charts identify unstable points or trends (special causes) for you by turning them red. Once you have selected your KPIs, you will want to track the stability and capability of the processes that produce them.

Choosing the Right Control Chart for Sustaining Improvement

One area that most people seem to struggle with is choosing the right control chart. The right chart is based on the type of data—counted or measured—and the sample size. The Control Chart Wizard in the QI Macros will analyze your data and select the right control chart for you. To choose the right chart yourself, follow these guidelines.

Choosing a Control Chart

With the recently added I-MR-R chart, there are now 14 control charts in the QI Macros. How do you know which one to use? When I'm working with data in Excel, I follow a simple strategy for selecting the right chart based on the *format* of the data itself. There are three formats I look for:

- A single row/column
- Two rows/columns with a numerator and a denominator
- Two or more rows/columns containing multiple measurements

Single Row/Column

If you only have a single row/column of data, there are only two main charts you will use:

- *C chart (attribute or counted data).* The data is always an integer (e.g., 1, 2, 3, 4, 5). For examples, look in QI Macros Test Data/cchart.xls.

- *XmR chart (variable or measured data).* The data usually has decimal places (e.g., 33.75). For examples, look in QI Macros Test Data/ xmrchart.xls.

So which one should you choose? If you're counting indivisible things such as defects, people, cars, or injuries, then choose the *c* chart. If you're measuring things such as time, length, weight, or volume, choose the XmR chart. If you're using a ratio such as defects per million parts, use the XmR chart. Look for these patterns in the data, and then select the chart.

Two Rows/Columns

If the data has a numerator and a denominator (e.g., defects per batch, errors per transaction), then you will want to use the:

- *P chart (parts are either good or bad).* For examples, look in QI Macros Test Data/pchart.xls.
- *U chart (one or more defects per piece).* For examples, look in QI Macros Test Data/uchart.xls.

How can you tell which one to use? I ask myself, "Can this widget have more than one defect?" If yes, use the *u* chart; otherwise, use the *p* chart.

Sometimes you can have more defects than samples (e.g., patient falls per 1,000 patient days). This is another clue (use a *u* chart). Again, look for these patterns in the data, and then select the chart. You can also convert these into a ratio (numerator/denominator) and use the XmR chart.

Two or More Rows/Columns of Variable Data

Service industries don't use these charts very often. They are used mainly in manufacturing when sampling two or more parts. If you have two or more rows or columns of variable data (e.g., time, weight, length, width, diameter, or volume), then you can choose one of four charts: XbarR, XbarS, XmedianR, or I-MR-R. For automotive examples, look in QI Macros Test Data/AIAG SPC.xls.

- XbarR (average and range, 2 to 10 rows/columns per sample)
- XMedianR (median and range, 2 to 10 rows/columns per sample)
- XbarS (average and standard deviation, 5 to 50 rows/columns per sample)
- I-MR-R [average, moving range (between subgroups), and range (within subgroups), 2 to 50 rows/columns per sample]

	A	B	C	D	E	F
7	Diameter 1	Diameter 2	Diameter 3	Diameter 4	Diameter 5	
8	22.30	22.54	22.01	22.62	22.65	
9	22.86	22.68	22.43	22.58	22.73	
10	22.88	22.68	22.46	22.30	22.61	USL
11	22.44	22.66	22.48	22.37	22.56	23.5
12	22.59	22.65	22.78	22.58	22.33	LSL
13	22.37	22.34	22.75	22.71	22.51	21.5
14	22.23	22.36	22.90	22.45	22.48	
15	22.60	22.72	22.35	22.51	22.69	
16	22.61	22.52	22.52	22.49	22.31	
17	22.42	22.64	22.52	22.40	22.63	

Figure 10.1 XbarR data.

The data should look like Figure 10.1.

You can run the XbarR, XMedianR, XbarS, or I-MR-R charts on the data in Figure 10.1. Xbar uses the average as the measure of central tendency. XMedianR uses the median. If you have more than five samples per period, then XbarS will probably be the most robust chart for your needs. You can also use the XbarR or XbarS if your data has a varying number of samples per period. The I-MR-R chart is like a combination of an XbarR and XmR chart; it measures the variation *within* subgroups with the range chart and variation *between* subgroups using the moving-range chart. Again, look for these patterns in your data, and then select the chart.

The np Chart

There's one chart that I've left to last because I rarely find situations where it applies. The *np* chart is like the *p* chart except that the sample sizes are constant. In business, sample sizes are rarely constant. The data looks like Figure 10.2. For examples, look in QI Macros Test Data/npchart.xls.

Download the QI Macros Control Chart Quick Reference Card (www.qimacros.com/sustainaid.pdf) for a control chart selection guide.

	A	B	C	D	E	F	G	H	I	J	K	L	M	N	O	P	Q	R	S	T	U	V	W	X	Y	Z
9	Spot Weld Flaws	2	5	4	3	3	6	5	0	7	5	4	1	2	3	6	3	8	4	4	4	6	4	2	3	7
10	Sample	62	62	62	62	62	62	62	62	62	62	62	62	62	62	62	62	62	62	62	62	62	62	62	62	62
11																										

Figure 10.2 *Np* chart data.

Other Control Charts

Once you understand how one control chart works, it's easy to understand the others. There are many other forms of control charts for various applications: short-run charts, analysis-of-means (ANOM) charts, exponentially weighted moving-average (EWMA) charts, moving-average charts, Levey-Jennings charts, and Hotelling charts.

- *Short-run charts (DNOM).* What if you only make three of this product and five of that one? There's never enough data to do a full control chart. Short-run charts analyze the deviation from nominal (target) for each different product.
- *ANOM charts.* ANalysis-Of-Means (ANOM) control chart shows variation from the mean. It's used mainly for experimental, not production, data.
- *CUSUM charts.* A CUmulative SUM (CUSUM) control chart detects small process shifts by analyzing deviation from a target value.
- *EWMA charts.* An exponentially weighted Moving-Average [a.k.a. geometric moving-average (GMA)] chart is effective at detecting small process shifts but not as effective as *X* charts for detecting large process shifts.
- *Moving-average charts.* A moving-average chart can be more effective at detecting small process shifts than an XmR chart. An EWMA chart may be more effective than a moving-average chart.
- *Levey-Jennings charts.* An average and standard-deviation chart is used extensively in laboratories.
- *Hotelling charts.* What do you do if you need to control two things simultaneously, such as vertical and horizontal placement of a drilled hole? Hotelling charts will assist in controlling these *multivariate* kinds of situations.
- *G and t charts.* Geometric-median and time-between control charts are used for rare events such as wrong-site or wrong-patient surgeries in a hospital. Hospitals use these charts to track *never events*—things that should *never* happen but do.

I have yet to use any of these charts in standard practice, but obviously, some people have advanced applications for them. I recommend getting familiar with *X*, *c*, *p*, and *u* charts before turning to these other types of charts.

If you learn to look for these patterns in your data, it will make it easier to choose the right control chart. Or you can let the QI Macros Control Chart Wizard pick them for you. And it's so easy to draw these charts with the QI Macros that you can draw them and throw them away if they aren't quite right.

Stability Analysis

A stable process produces *predictable results consistently*. Predictability can be easily determined using control charts. The control chart used to identify and focus your problem is a line graph with average and limit lines to help determine stability. The average and limit lines (±1, 2 and 3σ) are calculated from the data by the QI Macros. The *upper control limit* (UCL) is the +3σ line, and the *lower control limit* (LCL) is the −3σ line; 99.7 percent of all data points will fall between these two limits. This means that only 3 points out of 1,000 (0.03 percent) would normally fall outside the UCL or LCL.

Once you've got a control chart, what do you do? Processes that are out of control need to be stabilized before they can be improved. Special causes require immediate cause-effect analysis to eliminate variation.

The diagram in Figure 10.3 will help you evaluate stability in any control chart. Each of these conditions should occur no more than 0.03

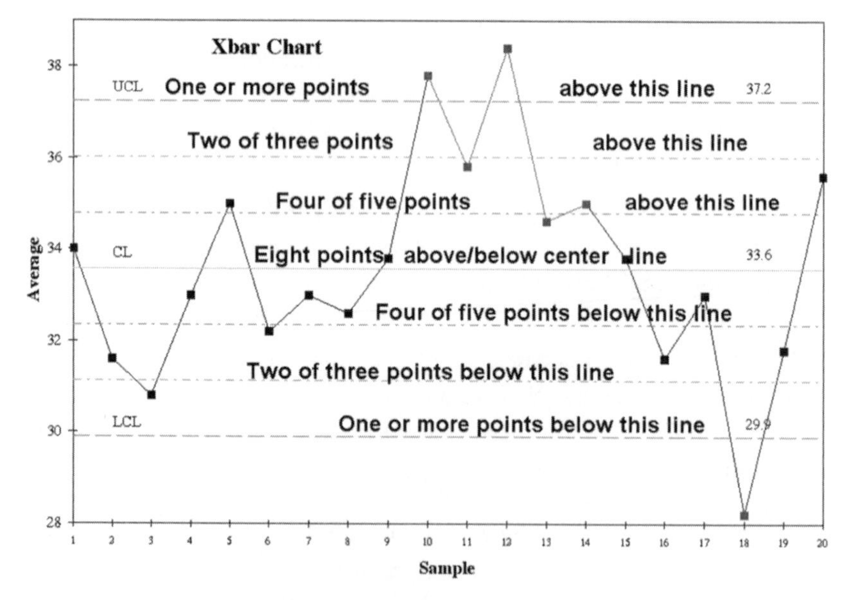

Figure 10.3 Stability analysis.

percent of the time, so to find any of them on a chart of 20 to 50 data points is highly unlikely. The QI Macros will pinpoint these for you.

Unstable conditions can be any of the following:

- Any point above the UCL or below the LCL
- Two of three points between 2σ and the control limits
- Four out of five points between 1σ and 2σ
- Eight points in a row above or below the center line
- Six points in a row ascending or descending (i.e., a trend)
- And there are several other rules, called *Nelson rules*, that detect other statistically unlikely conditions.

The QI Macros will automatically identify any violation of these rules by turning points and lines red.

Any of these conditions suggests that an unstable condition may exist. Investigate these special causes of variation with cause-effect analysis using the fishbone diagram or five whys (i.e., just ask "Why?" five times). Once you've eliminated the *special causes*, you can turn your attention to using the problem-solving process to reduce the *common causes* of variation.

Analyzing Capability

Although stability analysis detects how predictable a process might be, it still doesn't tell us whether the process delivers what the customer wants. A capable process *meets customer requirements 100 percent of the time.*

The capability target for counted data (e.g., defects) is zero. Customers hate defects, outages, and so on. How many wrong-site surgeries will patients tolerate? Zero, but there's usually about 30 or more per state each year. The capability of measured data such as time, money, age, length, weight, and so on is determined using the customer's specifications and a histogram.

When a customer defines an upper and a lower specification limit (USL and LSL) for a product or service, whether it's the diameter of a pipe or the time in line at a fast-food restaurant, all points within the two limits are considered "good." This involves understanding goalposts and targets.

Goalposts and Targets

The goal for all problems associated with variation is to center the distribution over the ideal target value and minimize the amount of variation around that target value. Sounds easy, doesn't it?

For most products, customers have a *target value* and some *tolerance* for variation around the target value. At my favorite restaurants, for example, I have a certain expectation for size, weight, and temperature of my favorite dishes. On any given day, they may be a little bigger or smaller, a little hotter or cooler, but if it's too cold, I might send it back. If the restaurant starts to skimp on portions, I might stop going. Your ability to produce products centered around the target value with a minimum amount of variation will affect the perceived quality of your product or service.

For parts to fit together properly, the bolt cannot be bigger or smaller than its nut it screws into; a cap cannot be bigger or smaller than its bottle. In many ways, this is like the goalposts in an American football game: there's a left and a right goalpost, and the kicker's job is to kick the ball between the two goalposts. Anything outside the goalposts results in no score. The left and right goalposts might be considered to be the game's *specification limits.*

TIP Don't confuse these *specification limits* (i.e., USL and LSL) with *control limits* (UCL and LCL). Customers set specification limits; QI Macros control charts use data to *calculate* control limits.

Customers specify their requirements for targets and tolerances in one of two ways:

- Target and tolerance (e.g., 74 ± 0.05)
- Upper (USL) and lower (LSL) specification limits (e.g., LSL = 73.95, USL = 74.05)

Piston heads, for example, will have specifications for the maximum and minimum height and diameter of the head, roundness of the head known as *cylindricity* and *concentricity*, and a host of other factors such as how the shaft connects to the piston and so on. A bottle will have similar specifications. Usually, a manufactured part will have both an upper and lower specification limit.

For most services, customers may have an upper limit but no lower limit. Teller wait times and call-center wait times usually will have only a maximum time (the minimum time is automatically zero). Most customers don't want to wait longer than 5 minutes in a bank and no more than 30 seconds on the phone. These customers have an upper specification limit but no lower limit other than zero. Go into any fast-food restaurant and you'll see a little digital timer ticking away next to your order. Fast-food restaurants can't afford to be slow because customers are paying for speed and convenience. Taco Bell, for example, aims to take and deliver a drive-through order in fewer than 180 seconds.

There are instances where you will have only a lower specification limit but no upper specification limit. The life of a light bulb, for example, might have a minimum of 1,000 hours but no maximum.

Distributions

It doesn't matter if you're measuring height, weight, width, diameter, thickness, volume, time, or money; if you measure the same dimension over time, it will produce a *distribution* that shows the variation. Most people have heard of a bell-shaped curve (Fig. 10.4); this is a *normal* distribution. Distributions have three key characteristics: center, spread, and shape (Fig. 10.5). The *center* is usually the *average* (a.k.a. the *mean*) of all the data points, although other measures of the center can be used [e.g., *median*

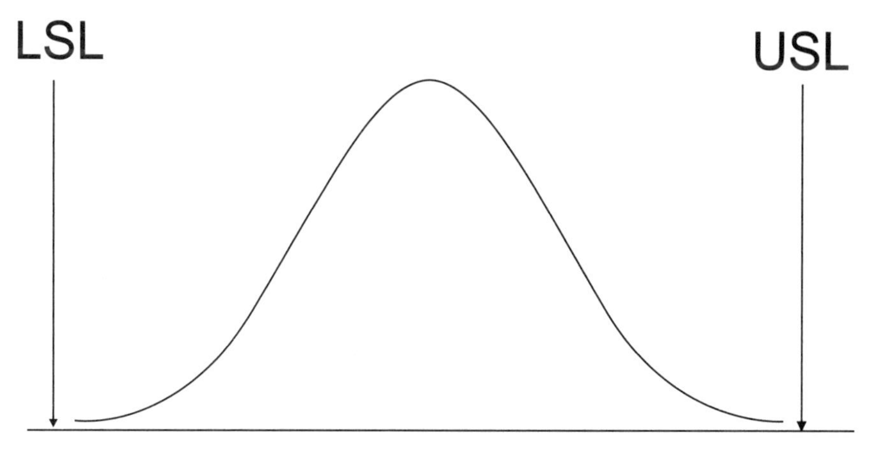

Figure 10.4 Bell-shaped curve.

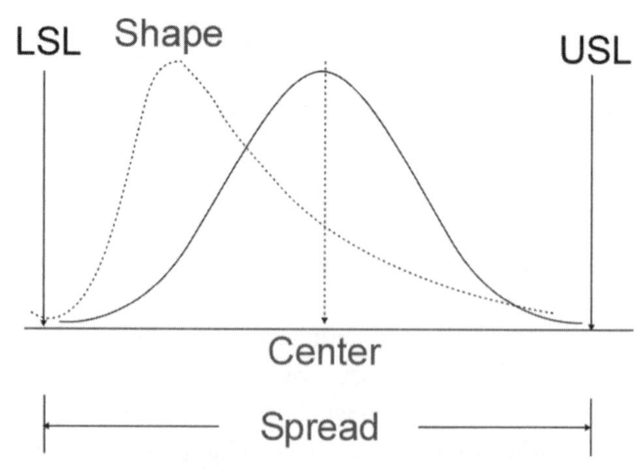

Figure 10.5 Center spread and shape.

(center point) or *mode* (most frequent data value)]. *Spread* is the distance between the minimum and maximum values. And the *shape* can be bell-shaped or skewed (i.e., leaning) left or right.

There are two outcomes for your improvement effort:

1. *Center* the distribution over the target value, as shown in Figure 10.6.
2. *Reduce the spread* of the distribution (i.e., reduce variation), as shown in Figure 10.7.

These two outcomes can be easily monitored using histograms, which help you to determine the capability of your process.

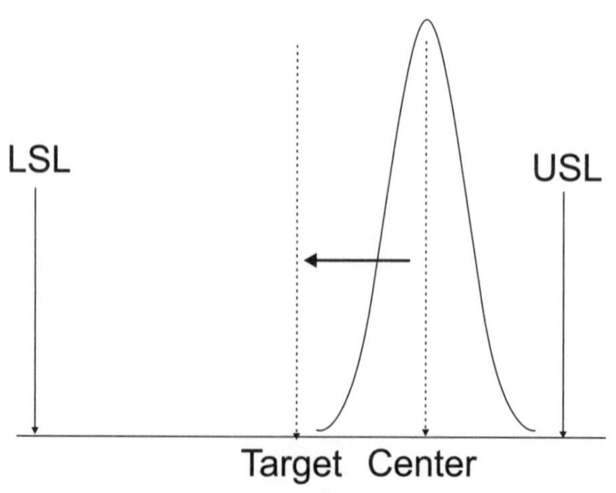

Figure 10.6 Center the distribution.

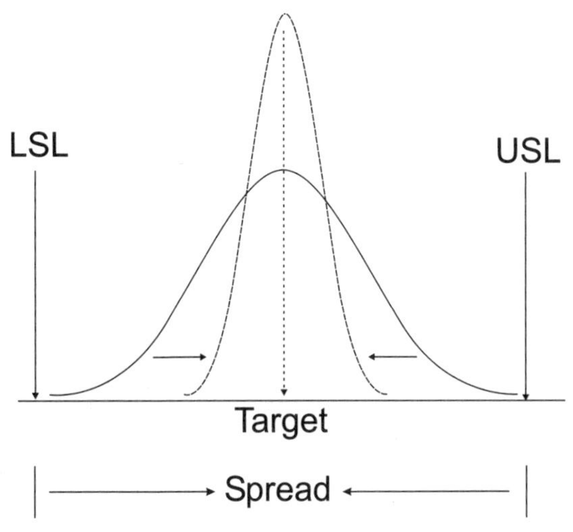

Figure 10.7 Reduce variation.

Histograms and Capability

Perhaps the easiest way to determine the center, spread, and shape of your data's distribution is with a histogram (Fig. 10.8). Histograms are simply bar charts that show the number of times your data points fall into each of

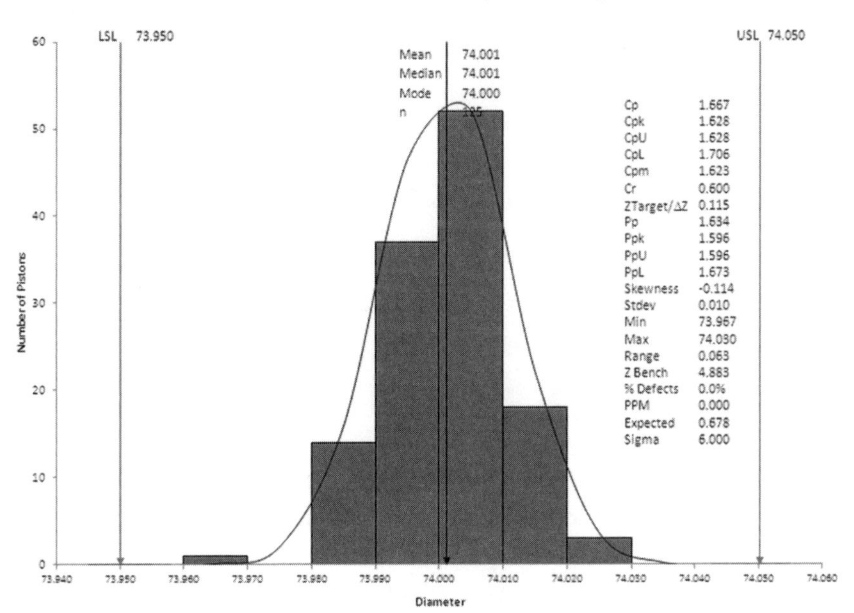

Figure 10.8 Histogram of piston head diameters.

the bars on the histogram. When you add the upper and lower specification limits, it's easy to see how your data fits your customer's requirements and what improvements might be necessary.

Capability Indices

Using the specification limits, there are four key indicators of process capability:

1. *Cp is the capability index.* It measures how well your sample fits between the upper and lower specification limits. It doesn't really care if the process is centered within the limits, only whether the data would fit if the data was centered. The histogram in Figure 10.8 has a *Cp* of 1.667, well above the minimum of 1.33.
2. *Cpk is the centering capability index.* It measures how well your sample is centered between the upper and lower specification limits. *Cp* and *Cpk* use an estimation of the standard deviation to calculate the spread of your data. If the variation between samples is small, *Cp* and *Cpk* are better predictors of capability. The histogram in Figure 10.8 has a *Cpk* of 1.628.
3. *Pp is the performance index.* Like *Cp*, it measures how well your data fits within the USL and LSL. Unlike *Cp*, *Pp* uses the actual standard deviation of your data, not the estimate. The histogram in Figure 10.8 has a *Pp* of 1.634. This is only slightly lower than *Cp*.
4. *Ppk is the performance centering index.* Like *Cpk*, it measures how well your data is centered between the USL and LSL. Again, *Ppk* uses the standard deviation to determine the spread of your data. The histogram in Figure 10.8 has a *Ppk* of 1.596.

TIP If *Pp* is significantly less than *Cp*, the process may be unstable or the data may have been sorted prior to creating the histogram.

If you want to dig into the formulas for these indicators, go to my website, www.qimacros.com/control-chart-formulas/cp-cpk-formula.

NOTE These indicators are only valid when your process is stable (i.e., in statistical process control). Remember, first stable, then capable.

Cp and *Cpk* should be used together to get a sense of process capability. Using *Pp* and *Ppk* will help to confirm process capability. Ideally, all four indicators should be greater than 1.33 (all data fits within specification limits and is centered on the target).

Turnaround-Time Histogram

Most service businesses will want to keep track of turnaround (or cycle) times. How long does it take for the customer to get what he or she wants? And customers can be internal or external ones. Figure 10.9 shows turnaround times for radiology reports with an upper specification limit of 24 hours.

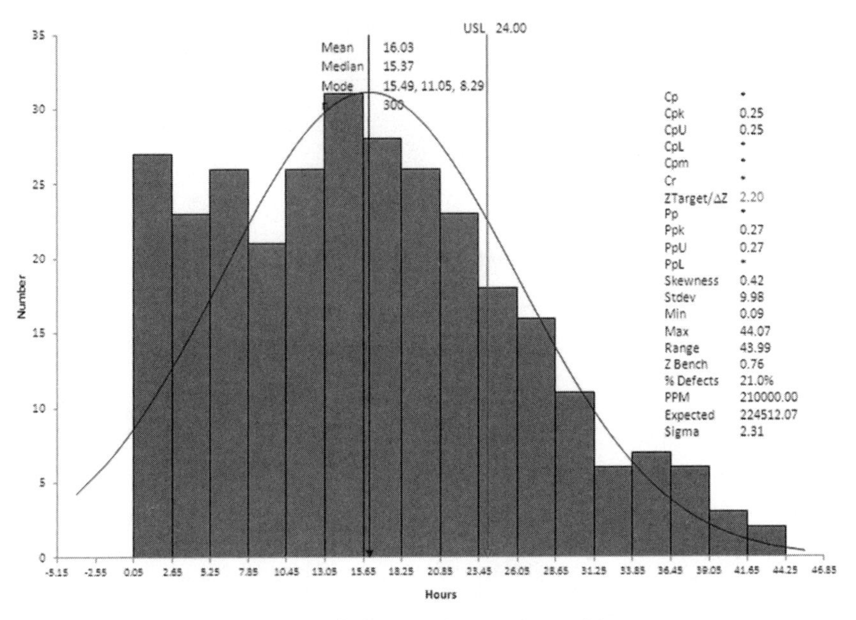

Figure 10.9 Radiology turnaround time histogram.

Because there is no lower specification limit, we can't calculate *Cp*, just *Cpk*. In this case, *Cpk* is only 0.25, which means that the process isn't centered. There are too many times far above the average (mean) of 16.03. This is the sort of problem that is easily solved using value-stream mapping and Excel's drawing tools (see Chap. 11).

Control Plan

For companies that need more rigor in sustaining improvement, consider implementing a control plan (Fig. 10.10). A *control plan* is a structured method for identifying, implementing, and monitoring process controls. A control plan describes what aspects of the process, from start to finish, will be kept in statistical process control, and it also describes the corrective actions needed to restore control. Process flowcharts and failure mode and effects analysis (FMEA) documents support the development of the control plan. The QI Macros include fill-in-the-Blank templates for flowcharts, FMEAs, and control plans.

	A	B	C	D	E	F	G	H	I	J	K	L	M
1								CONTROL PLAN					
2		Prototype	Prelaunch	X	Production	Key Contact/Phone				Date(Orig)		Date (Rev.)	
3	Control Plan Number		12345				Jay A.	Injection Svcs		2/26/2010			
4	Part Number/Latest Change		45678			Core Team				Customer Eng. Approval/Date			
5	Part Name/Description		Dashboard			Organization/Plant Approval/Date				Customer Quality Approval/Date (if Req'd)			
6	Organization/Plant		IS/Plant 1	Organization Code		Other Approval/Date (If Req'd)				Other Approval/Date (If Req'd)			
7				Characteristics				Methods					
8	Part/ Proces s Numbe r	Process Name/ Operation Description	Machine, Device, Jig, Tools, for Mfg.	No.	Product	Process	Special Char. Class	Product/Process Specification/ Tolerance	Evaluation/ Measurement Technique	Sample Size	Freq.	Control Method	Reaction Plan
9	1	Injection Molding	Mold 1	1	Mounting hole			burrs	Inspection	1	1	Inspection	Adjust/recheck
10								15mm+/- 1mm	Gage	5 pcs	1/hr	XbarR Chart	Adjust/recheck

Figure 10.10 Control plan.

The control plan for any part, assembly, or deliverable identifies

- All steps in the manufacturing or service process (e.g., injection molding)
- Any machines used in the manufacture or delivery (e.g., mold 1)
- Product characteristics to be controlled (e.g., mounting-hole burrs and diameters)
- Specifications and tolerances (e.g., 15 ± 1 mm)
- Techniques for measurement and evaluation (e.g., gauges)
- Sample size and frequency of measurement (e.g., 5/h)
- Control methods (e.g., inspection, XbarR chart, etc.)
- Reaction plan—what to do when the characteristic goes out of control (e.g., adjust, recheck, quarantine)

Although control plans are beyond the scope of this book, it's useful to know that there's more rigor available if needed. There's a checklist for developing control plans, FMEAs, and flowcharts in the QI Macros APQP Checklist template.

QI Macros for Monitoring Stability and Capability

The QI Macros have Fill-in-the-Blank templates for monitoring stability and capability (Fig. 10.11). Control charts and histograms can be found under the "SPC Charts" pull-down menu.

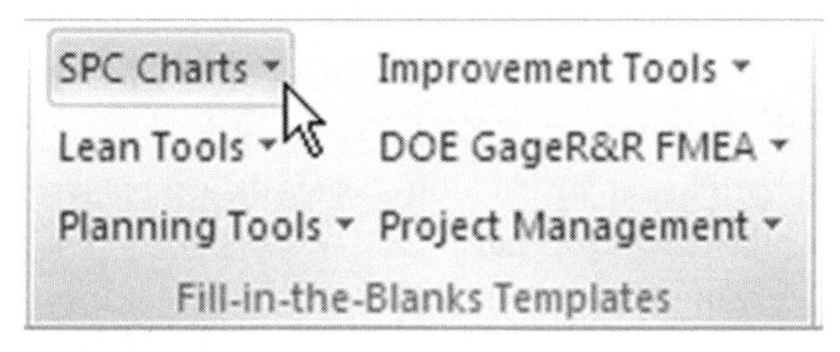

Figure 10.11 QI Macros Fill-in-the-Blank templates.

Create Control-Chart Dashboards in Minutes

There are also several control-chart dashboards to simplify monitoring many measurements at the same time (Fig. 10.12).

I have found that most companies spend too much time:

- Updating existing charts every month with new data
- Organizing multiple control charts onto a single page
- Creating control-chart dashboards for management

The QI Macros control-chart dashboards save you time by automating these tasks.

Figure 10.12 QI Macros control-chart dashboards.

Create an XmR Control-Chart Dashboard Using the QI Macros

View "Control Chart Dashboard Tutorial" (six minutes) at www.qimacros .com/training/qi-macros-webinar/qi-macros-dashboards, or follow these steps:

1. Click on the QI Macros pull-down menu, and select "Control Chart Dashboard." Then open the "XmR Dashboard."
2. Each dashboard has an instruction sheet, a data-input sheet, and a sheet for each available control chart.
3. Input or copy and paste your data into the "Data Input Sheet" (Fig. 10.13).

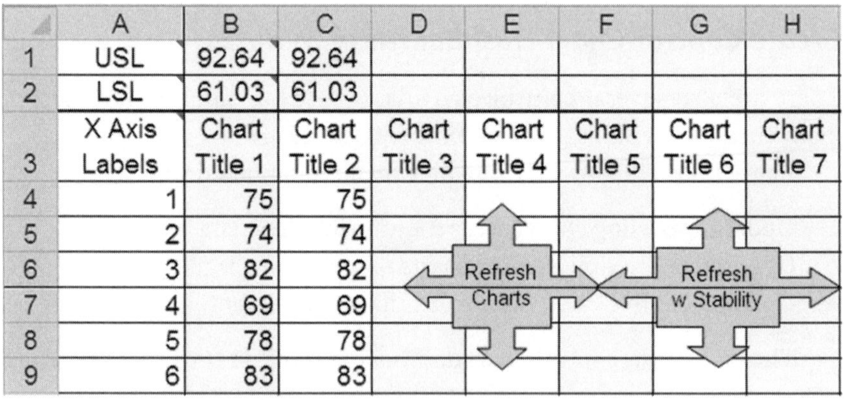

	A	B	C	D	E	F	G	H
1	USL	92.64	92.64					
2	LSL	61.03	61.03					
3	X Axis Labels	Chart Title 1	Chart Title 2	Chart Title 3	Chart Title 4	Chart Title 5	Chart Title 6	Chart Title 7
4	1	75	75					
5	2	74	74					
6	3	82	82					
7	4	69	69					
8	5	78	78					
9	6	83	83					

Figure 10.13 XmR control-chart dashboard.

NOTE ABOUT UPDATING CHARTS Whenever you add new data to this sheet, you can click on the "Refresh Charts" or "Refresh with Stability Analysis" buttons to update all the charts.

NOTE ABOUT CHART TITLES Because we use these chart titles to name sheets when the dashboard is created, make these names unique, and limit them to 31 characters.

XmR Dashboard Data Sheet

Rows 1 and 2 of the XmR dashboard data sheet contain input cells for USLs and LSLs. Input your specification limits here to be used in the histogram calculations. *Note:* LSL and USL defaults are estimated as the average $\pm 3\sigma$.

Now click on the sheet for the desired chart (Fig. 10.14). In the XmR dashboard, there are four chart choices:

- *XmR average*, which will handle up to 100 data points
- *XmR rolling*, which will show the most recent 50 data points
- *XmR median*, which uses the median as the center line
- *Run chart* (I don't recommend run charts, but some people love them)

Click on the previous and next arrows (*top right* in the figure) to scroll through charts for each column of data. Change the chart formatting as needed, and then click on the "Create Dashboard" button (*bottom right*) to create a dashboard with each chart.

Once you click on "Create Dashboard," a macro will run for several seconds and create a new dashboard sheet with all the *x* charts (Fig. 10.15).

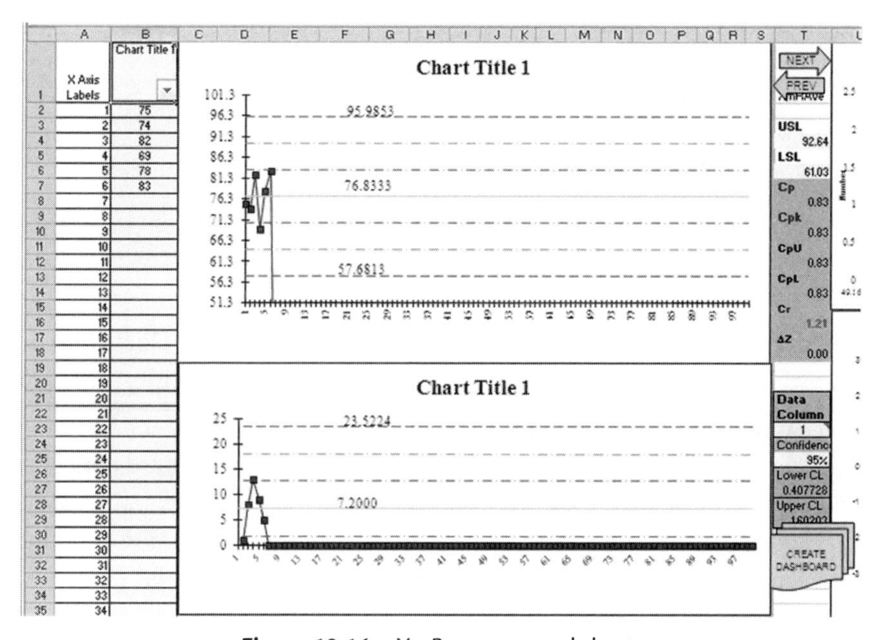

Figure 10.14 XmR average worksheet.

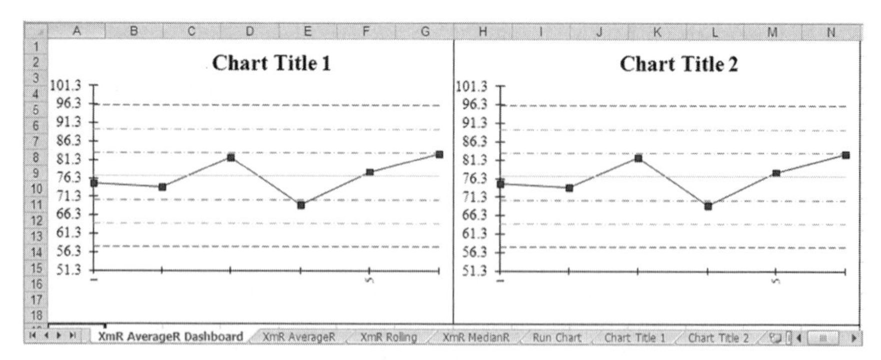

Figure 10.15 Control-chart dashboard.

To add new data to your charts, just go to the data-input sheet, and add your new data. The charts on the XmR or other control-chart templates will update automatically. To update the charts on the dashboards, click on the "Refresh Charts" icon. To update the charts and run stability analysis, click on the "Refresh with Stability" icon (Fig. 10.16).

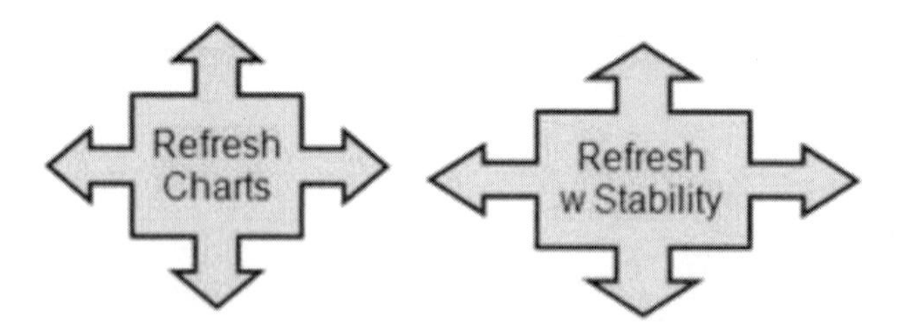

Figure 10.16 Refresh Charts and Refresh with Stability icons.

A Note about Adding New Charts After You Create a Dashboard

Once you create a dashboard, the number of charts on the dashboard is set. You cannot add data into a new chart column on the data sheet and add it to the existing dashboard. You will need to open a new dashboard template and cut and paste your data input sheet into the new template and create a new dashboard. As discussed earlier, you can always add data to existing charts; you just can't add a new chart to an existing dashboard.

Summary

Once you've achieved a breakthrough improvement, you can use the QI Macros control charts, histograms, and dashboards to monitor the process to make sure that the improvement sticks. Otherwise, the process will slowly slip back to prior levels of performance. It's not enough to improve; you also have to *sustain* the improvement. This is where most breakthrough improvements fail.

CHAPTER 11

Excel Drawing Tools for Breakthrough Improvement

Amazon ships products within 2.5 hours of a customer order, according to a September 2013 *Fast Company* article entitled, "The Race Has Just Begun." Although this packing and shipping process involves a lot of manual work, Amazon has cut the time to do it by 25 percent in the last two years. In some cities, orders can be delivered on the same day.

"Bezos has turned Amazon into an unprecedented speed demon that can give you anything you want. Right. Now." says J. J. McCorvey, the article's author. What would happen to your business if you cut the order-fulfillment process by 25 percent? Happier customers? Fewer errors? More sales?

Businesses worldwide are applying improvement methodologies to drive productivity and profitability. Experience has shown that as little as 4 percent of any business produces over half the delay, defects, and deviation. Finding and fixing these pockets of dysfunction can reduce costs and boost profits dramatically. According to authors George Stalk and Thomas Hout in *Competing Against Time*, a 25 percent reduction in delay *will double productivity and boost profits by 20 percent*. It doesn't matter if you're a manufacturer making parts or a health care provider treating a patient—focusing on removing delays can help you to fire up your profits and productivity.

Speed is the new killer app. Is your business up to Amazon speed?

Visual Thinking for Breakthrough Improvement

> If you want everyone to have the same mental model of a problem, the fastest way to do it is with a picture.
>
> —DAVID SIBBET

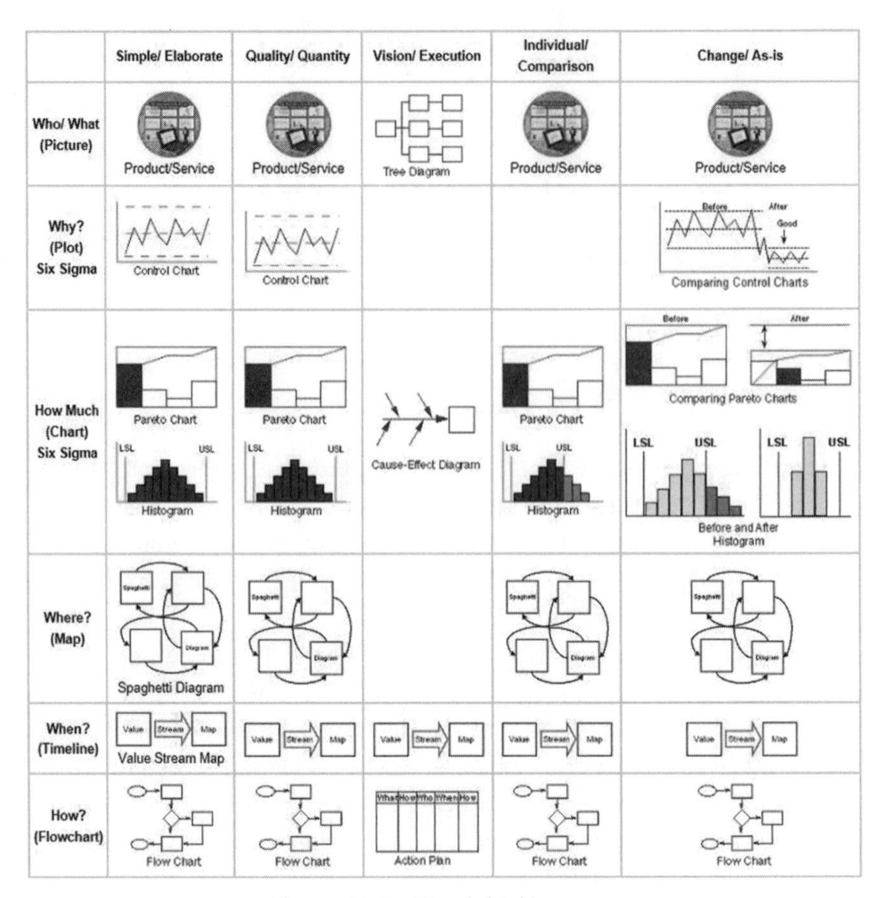

Figure 11.1 Visual thinking map.

Dan Roam created a tool for visual thinking and problem solving that applies directly to breakthrough improvement. I've modified his map (www.napkinacademy.com) to use breakthrough improvement tools and methods (Fig. 11.1).

Problem Solving Is a Story Told with Pictures

The average manager will look at a spreadsheet for only a few seconds. Turn the numbers into pictures that convey all the information needed in a handful of charts and diagrams that can be grasped easily.

When it comes to problem solving, we need to answer Who? Why? How much? Where? When? and How? And we can do it with pictures, charts, and diagrams.

- *Why?* Control chart of current performance (e.g., defects, deviation, delays)
- *How much?* Pareto chart of defects or histogram of deviation
- *Where?* Spaghetti diagram of workspace and product flow
- *When?* Value-stream map of process flow and delays
- *How?* Flowchart of process, action plans for improvement, and control plan

These last three—Where? When? and How?—can be drawn with Excel's drawing tools (Fig. 11.2). Most people think that they need Visio to draw these kinds of diagrams. Not true. Excel will do everything you need. So how do we start to achieve Amazon speed? By learning how to *mind the gap*.

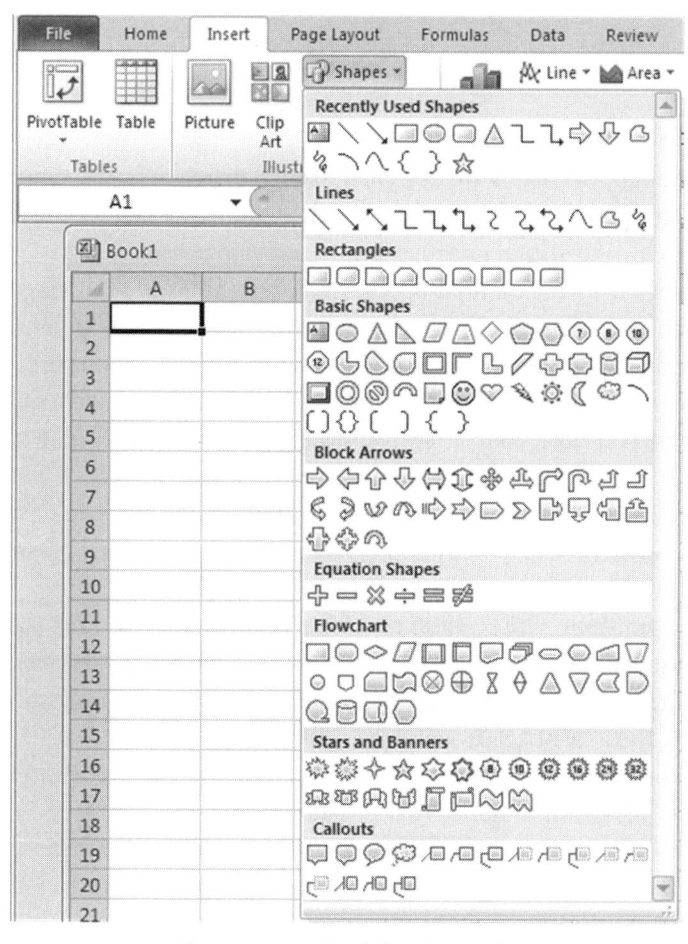

Figure 11.2 Excel drawing tools.

Mind the Gap

If you've ever been to London and ridden on the famous Underground, you've probably seen signs that say, "Mind the Gap" (Fig. 11.3).

Figure 11.3 Mind the gap.

Although such signs are designed to keep travelers from wrenching an ankle in the gap between the train and the platform, I believe the idea also applies to breakthrough improvement.

Here's what I mean: hold up your hand, and spread your fingers wide apart. What do you see? Most likely you're first drawn to look at your fingers, not the gaps in between. This is how most people think about speeding up their business—by looking at the employees, not at the *gaps between people.*

When you take your attention off the employees and put it on the product or service going through the process, you quickly discover that there are huge gaps between one step in the process and the next. You'll discover work products piling up between steps, which only creates more delay—a bigger gap.

Your problems with sluggish performance are in the gaps, not the people. You can make the people work faster, but you'll find that this often makes the business slower, not faster, because more work piles up between steps, widening the gap, not narrowing it.

The other sign that you often see in the London Underground is a tube map (Fig. 11.4). You'll notice that the stations are quite small and the lines between them quite long. This is true of most processes. The distance between workstations is greater than necessary. If you want to reduce the time it takes to serve a customer, you have to mind the gaps.

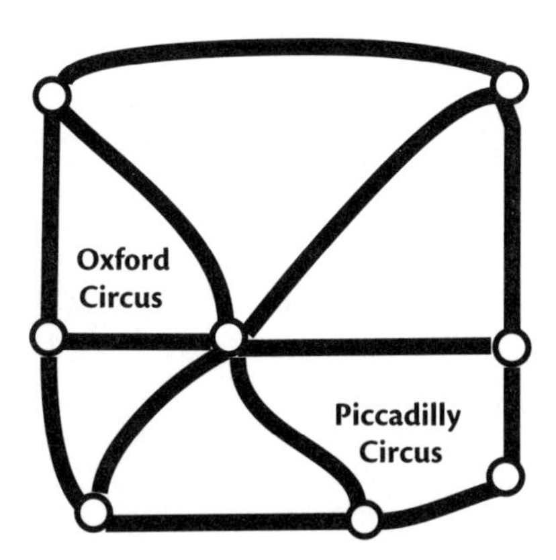

Figure 11.4 Tube map.

Most businesses have a blind spot when it comes to the best way to accelerate their response time. Stop trying to make your people faster. Stop trying to keep your people busy when there's no reason to be busy. Stop making things customers haven't yet asked for. Stop making people and materials travel too far to get the job done. To get up to Amazon speed, focus on the delays and movement between steps, not the steps themselves.

When you do this, you will get an added benefit: 50 percent fewer defects, mistakes, errors, and glitches. When you stop picking the product or service up and setting it down and picking it up and setting it down, you reduce the opportunity to make a mistake or miss a step. You get better products or services in half the time.

Breakthrough Improvements in Speed

A breakthrough improvement project to increase speed often begins with no more than a few pads of Post-It Notes to map the flow of a process. A flowchart uses a few simple symbols to show the flow of a process: ovals, boxes, diamonds, and arrows (Fig. 11.5).

- *Ovals.* These represent the starting and ending points from the customer's point of view (e.g., customer places order, customer pays bill for shipped product, etc.).
- *Boxes.* These represent activities that add value to the product or service. These should be indicated as a verb-noun (e.g., fill order, calculate bill, post payment, etc.).

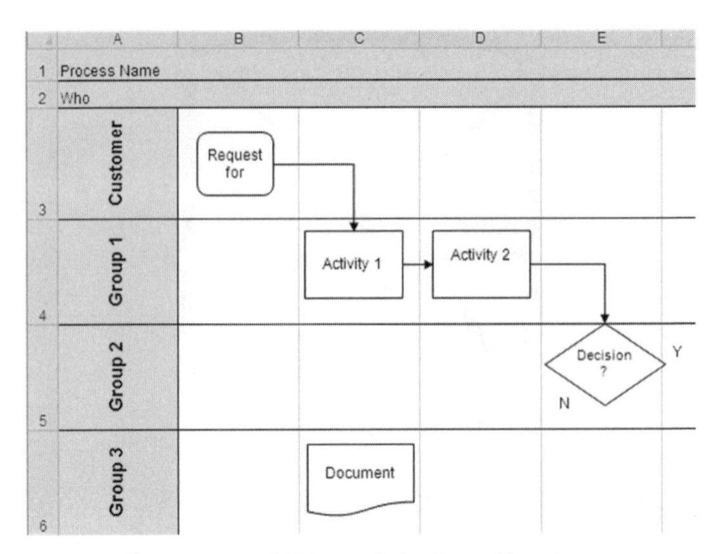

Figure 11.5 QI Macros Swim Lanes Flowchart.

- *Diamonds.* These represent decisions where someone is choosing between two or more alternatives (e.g., order approved, e-mail or snail mail, etc.).
- *Arrows.* These represent the flow of the process and the transition between activities or decisions. (I found sticky notes shaped like arrows on Amazon.)
- *Circles.* These represent where to measure the quality or process indicators (usually at decision points).

How to Develop a Flowchart

Use square Post-It Notes for both the decisions (diamonds) and activities (boxes). In this way, the process will remain easy to change until you have it clearly and totally defined. Limit the number of decisions and activities per page. Move detailed subprocesses onto additional pages. Across the top of the flowchart, list every person or department that helps to deliver the product or service. Along the left-hand side, list the major steps in your process: planning, doing, checking, and acting to improve. Even going to the grocery store involves creating a list (plan), getting the groceries (do), checking the list (check), and acting to get any forgotten items.

Using the QI Macros Swim Lanes Flowchart, you can then capture the flow into Excel, where it can be modified and improved as required.

Value-Stream Maps for Breakthrough Improvements in Speed

I have worked with government agencies that found ways to take 110 days out of a 140-day process and corporate teams that took 8 hours out of a 9-hour process. It doesn't matter how long a process seems to take. A team usually can figure out how to cut the time in half after a few hours of analysis using value-stream mapping (VSM). Value-stream maps are especially helpful for removing unnecessary *delay*. Check out these how-to videos at www.breakthrough-improvement-excel.com.

Breakthrough improvement using a value-stream map (Fig. 11.6, QI Macros "Value-Stream Mapping" template) is easy:

- Start with the high-level process steps.
- Use Post-It Notes to lay out steps (squares) and delays between steps (arrows).
- Put estimated times on each step and arrow.
- Identify the biggest delays between steps.
- Figure out how to redesign the process to (1) eliminate delays or (2) do two or more steps simultaneously.

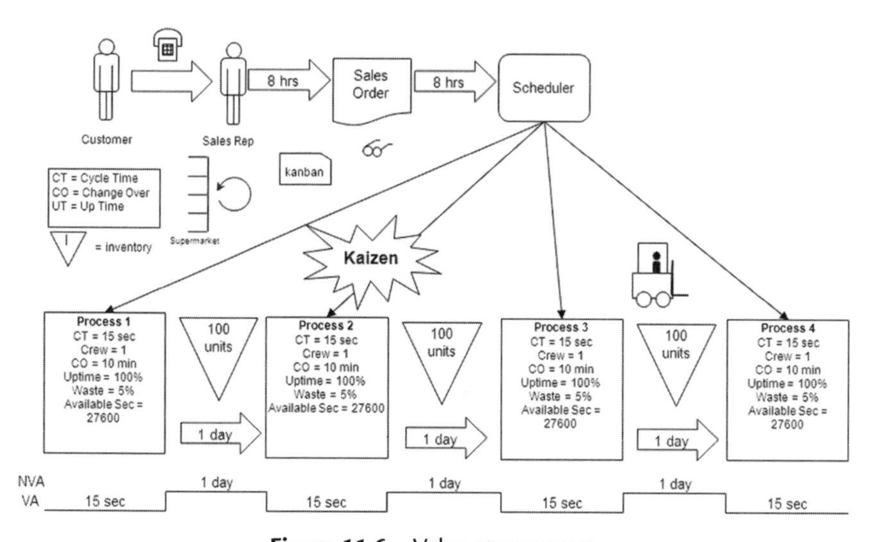

Figure 11.6 Value-stream map.

TIP Use Post-It Notes to draft the flow. I have found that few people have the same mental map of the flow, so it can take some time and tweaking to settle on the actual flow. Then, and only then, capture it with Excel's drawing tools and the QI Macros "Value-Stream Mapping" template. There are companies that make arrow-shaped sticky notes (Reditag 41254), but you can also just draw an arrow on a square Post-It.

HINT Ninety percent of the delays are in the arrows *between steps.*

Breakthrough Improvements by Eliminating Movement

Amazon doesn't keep all its books in one place and kitchen equipment in another. Baking cookbooks and mixers might be next to each other because customers buy them together. This prevents unnecessary movement.

When I go into my local grocery store to buy coffee, I'm always surprised to find that coffee filters may be located aisles away in paper products. (Sounds silly, doesn't it?) I discovered that my local Ace Hardware keeps the coffee filters right by the coffee machines. It's often easier for me to pick up filters at Ace than to hunt them down at the grocery store.

While value-stream maps and flowcharts can help us to find and fix delays, we need another kind of diagram to reduce unnecessary movement of people and materials. The aptly named *spaghetti diagram* (Fig. 11.7) helps show the physical movement of customers, employees, and work. Spaghetti diagrams are especially useful for eliminating unnecessary movement of people, machines, or materials.

To create a spaghetti diagram:

1. Start with a piece of paper to represent the workspace. Use Post-It Notes to diagram the workstations or machines used.
2. Draw arrows to show the movement of a product through the workspace. If you're not sure how it moves, follow one around.
3. Count how many times each machine or workstation is used. Sometimes the highest-volume location is farthest from the incoming work.
4. Rearrange the machines, materials, or work areas to reduce movement of people and materials.

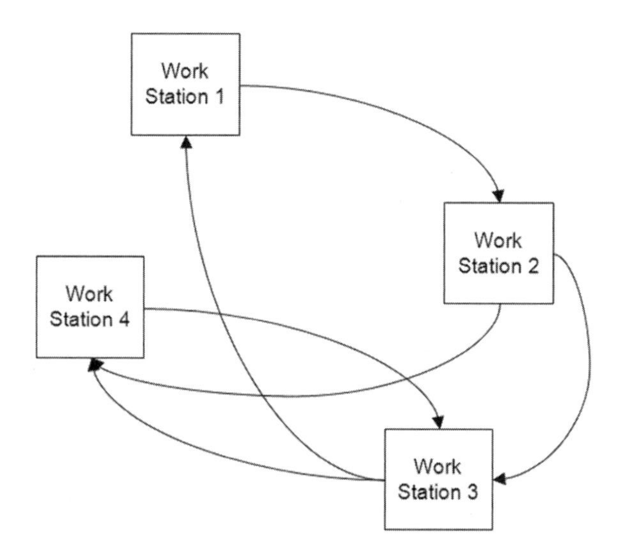

Figure 11.7 Spaghetti diagram.

5. Look for opportunities to reduce or eliminate:
   - Delay
   - Overproduction
   - Waste and rework (defects)
   - Non-value-added processing
   - Transportation (unnecessary)
   - Inventory
   - Motion (unnecessary)

Walking Is Waste!

You can use pedometers to measure how far employees are traveling on a daily basis. A friend of mine started working in a new hospital clinic. Using pedometers, management quickly discovered that nurses were walking *10 to 12 miles a day* because of how the hospital was designed. Simple redesign of the machines and materials slashed this to a fraction of the previous level.

Using value-stream maps and spaghetti diagrams, it is not uncommon for teams to quickly find ways to reduce turnaround time by 50 to 90 percent. And that's what customers want, isn't it? Faster service. Even faster tomorrow.

Time-Tracking Template

Sometimes you will need to track the times for a process. The QI Macros "Time-Tracking" template (Fig. 11.8) will help to automate the collection of data for the time between steps within a process. Because most people struggle with how to input dates and times into Excel, I've created a macro to do it for you. Just click on the cells in columns A, C, and D, and the template will populate the cell with the current time and date. The template has worksheets for up to 10 steps.

	A	B	C	D	E	F	G	H	I
			Start	Times					
1	Date	Ref #	Step 1	End Time	Who?	Reason for Delay	Total Time Start to End hh:mm	Total Time Start to End Minutes	Total Time in Hours
2	05/18/09	409	09:33	09:53	JS		0:20	20	0.33
3	05/19/09	123	11:30	12:05	LJ		0:35	35	0.58
4									
5									
6									

Figure 11.8 Time-Tracking template.

To use the Time-Tracking template:

1. Click on the cell in column A to populate the current date.
2. Click on the cell in column C when the process starts to populate the date and time.
3. Click on the cell in column D when the process stops to populate the date and time.
4. Populate columns B, E, and F with other information about each event.
5. Column G is calculated as the difference between the start and stop times (columns C and D).
6. Column H converts column G into minutes. This conversion is usually necessary to run charts.
7. Select the data in column H to run a control chart.
8. To edit dates or times in columns A, C, and D, click on the cell you want to edit, and make the changes.

Value-stream maps and spaghetti diagrams are the easiest way to start achieving breakthrough improvements in speed, and they also reduce errors. You can use Excel's drawing tools and the QI Macros Fill-in-the-Blank templates to capture the current and improved processes.

Breakthrough Improvement Exercises

Use Post-It Notes to develop:

- *A value-stream map of one of the key processes in your workplace.* How can you eliminate delays or do steps in parallel to reduce cycle time?
- *A spaghetti diagram of how a product or service flows through your workplace.* How can you rearrange the workspace and materials to eliminate unnecessary movement?

CHAPTER 12

Hypothesis Testing for Breakthrough Improvement

Some tools of breakthrough improvement aren't visual; they're simply analytical. Sometimes you want to be able to compare two processes or products and learn something about their quality using statistics alone. Health care research often uses these tools to compare the effects of two or more medications or protocols. In manufacturing, these statistical tools help to evaluate the various formulas or mixtures. You can use these tools to compare performance before and after an improvement to verify its effectiveness. This falls under the category of *hypothesis testing*.

Hypothesis Testing

I've come to suspect that hypothesis testing is where statistics got the nickname "sadistics." I found it confusing because it seems to use negative logic to describe everything. But it's really not that hard once you understand how it works, and the QI Macros make it easy to compare two or more sets of results.

Let's say that you have two medications or mixtures and you want to prove that they are (1) the *same* (i.e., equal) or (2) *different* (i.e., not equal) at a certain level of confidence. You might want to compare a control group with a test group to determine whether a medication is effective. You might want to compare your current formula for aluminum foil (an alloy of aluminum and other minerals) with a new formulation that is more affordable or stronger or whatever. You will want to know whether the averages (i.e., means) or variation are the same or different. Hypothesis testing helps you to evaluate these two *hypotheses—sameness* or *difference*.

There are several tools that can help you to do this depending on whether you are most interested in the average or the variability.

Statistics with QI Macros and Excel

You can install Excel's Data Analysis Toolpak to perform many of the tests in this chapter. Unfortunately, Excel doesn't really tell you what the results mean. It usually just gives you the probability (p-value) that the data is the same and makes you decide what it means.

RULE If the p-value is less than the significance level (typically 0.05), then the data isn't likely to be the same.

For statisticians, this is easy; for typical people who use statistics infrequently, it's hard to remember what to do. This is why the QI Macros compare the p-value with the significance level and tell you what it means in plain English. This makes it a lot easier to make decisions.

Hypothesis Testing for Variation

Because variation (i.e., too big/too small, too heavy/too light, etc.) can affect profits and productivity, it's useful to determine whether variation in two or more samples is the same or different. To evaluate *variation*, use the F-test or Levene's test.

Levene's Test for Variation

Levene's test helps to determine whether two or more variances are the same or different from each other. Let's suppose that you make paper and want to measure the effects of several hardwood concentrations on the tensile strength of the paper produced. Hardwood is more expensive than other ingredients, so you'd like to use the minimum required to get the desired strength. Let's suppose that your customer wants a tensile strength of at least 15 pounds per square inch. So you decide to test at four different concentrations—5, 10, 15, and 20 percent hardwood in the paper.

After conducting a test and entering the data into Excel (Fig. 12.1, QI Macros Test Data/anova.xls), use the QI Macros to conduct Levene's test. The QI Macros will prompt for a significance level (a typical default is 0.05).

	A	B	C	D	E	F
1	Hardwood Concentration %	5%	10%	15%	20%	Tensile Strength of Paper (PSI)
2	Obs1	7	12	14	19	
3	Obs2	8	17	18	25	
4	Obs3	15	13	19	22	
5	Obs4	11	18	17	23	
6	Obs5	9	19	16	18	
7	Obs6	10	15	18	20	

Figure 12.1 Tensile strength test data.

	A	B	C	D	E	F	G	H	I	J	K	L	M	N
1	5%	10%	15%	20%		5%	10%	15%	20%					
2	7	12	14	19	Median	9.5	16	17.5	21					
3	8	17	18	25	Mean	10	15.67	17	21.2					
4	15	13	19	22	Variance	8	7.867	3.2	6.97					
5	11	18	17	23	n	6	6	6	6					
6	9	19	16	18	df	5	5	5	5					
7	10	15	18	20		Levene's								
8					Test	0.599								
9					*p*	0.623		Accept Null Hypothesis because p > 0.05 (Variances are the same)						
10						a	0.05							

Figure 12.2 Levene's test results.

Then the QI Macros will calculate the results (Fig. 12.2). (*Note:* Levene's test is not part of Excel's Data Analysis Toolpak.)

Because the *p*-value 0.623 is greater than 0.05, we accept the hypothesis that variances are equal for different concentrations of hardwood.

F-Test for Variation

If you only want to compare two levels (e.g., 10 and 15 percent), use the *F*-test. It helps to determine whether the variances are the same or different from each other.

The QI Macros F-test (Fig. 12.3) will compare the *p*-value with the significance level (i.e., 0 .173 > 0.05). It is, so this means that the variances are statistically the same at a 0.05 level of significance. This tells us that the amount of hardwood doesn't affect tensile strength *variability*.

	A	B	C	D	E	F	G	H	I
1	10%	15%	F-Test Two-Sample for Variances	α	0.05				
2	12	14							
3	17	18		10%	15%				
4	13	19	Mean	15.67	17				
5	18	17	Variance	7.867	3.2				
6	19	16	Observations	6	6				
7	15	18	df	5	5				
8			F	2.46					
9			P(F<=f) one-tail	0.173	0.346	Two-tail			
10			F Critical one-tail	5.05	7.15	Two-tail			
11			One-tail	Accept Null Hypothesis because p > 0.05 (Variances are the same)					
12			Two-tail	Accept Null Hypothesis because p > 0.05 (Variances are the same)					

Figure 12.3 *F*-test results.

	A	B	C	D	E	F	G	H	I	J	K
1	10%	15%		10%	15%						
2	12	14	Median	16	17.5						
3	17	18	Mean	15.6667	17						
4	13	19	Variance	7.86667	3.2						
5	18	17	n	6	6						
6	19	16	df	5	5						
7	15	18		Levene's							
8			Test	2.118							
9			*p*	0.176		Accept Null Hypothesis because p > 0.05 (Variances are the same)					
10			F-Test	*a*	0.05						
11			F	2.46							
12			*p* 1&2 tail	0.173	0.346						
13			F Critical	5.05							

Figure 12.4 Levene's test results—two factors.

We could also use Levene's test on two samples (Fig. 12.4). Because Levene's *p*-value is greater (i.e., $0.176 > 0.05$), we can accept the null hypothesis that the variances are equal.

Hypothesis Testing for Means

Because the mean (i.e., average) also affects results, it can be useful to evaluate whether the means of one or more samples are the same or different. There are a number of ways to do this depending on sample size: analysis of variance (ANOVA), *t*-tests, and the Tukey test.

Analysis of Variance

Analysis of variance (ANOVA) can help you to determine whether two *or more* samples have the same mean or average. Excel and the QI Macros can perform both single- and two-factor ANOVA.

Single-Factor Analysis

Now we want to find out how the four different hardwood levels affect average tensile strength. Using the QI Macros, select the data, and use "QI Macros–Statistical Tools–ANOVA Single Factor" to run a single-factor ANOVA (Fig. 12.5).

	A	B	C	D	E
1	Hardwood Concentration %	5%	10%	15%	20%
2	Obs1	7	12	14	19
3	Obs2	8	17	18	25
4	Obs3	15	13	19	22
5	Obs4	11	18	17	23
6	Obs5	9	19	16	18
7	Obs6	10	15	18	20

G	H	I	J	K	L	M
Anova: Single Factor	α	0.05				
SUMMARY						
Groups	*Count*	*Sum*	*Average*	*Variance*		
5%	6	60	10	8		
10%	6	94	15.6667	7.86667		
15%	6	102	17	3.2		
20%	6	127	21.1667	6.96667		
ANOVA		Reject Null Hypothesis because p < 0.05 (Means are Different)				
Source of Variation	*SS*	*df*	*MS*	*F*	*P-Value*	*F crit*
Between Groups	382.8	3	127.597	19.6052	0.000	3.098391
Within Groups	130.2	20	6.50833			
Total	513	23				

Figure 12.5 Anova results.

Because the p-value is less than alpha (i.e., $0.00 < 0.05$), we can say that the means *are* different. Just for fun, you might want to run a box and whisker chart on the data to see the variation (Fig. 12.6). The variation

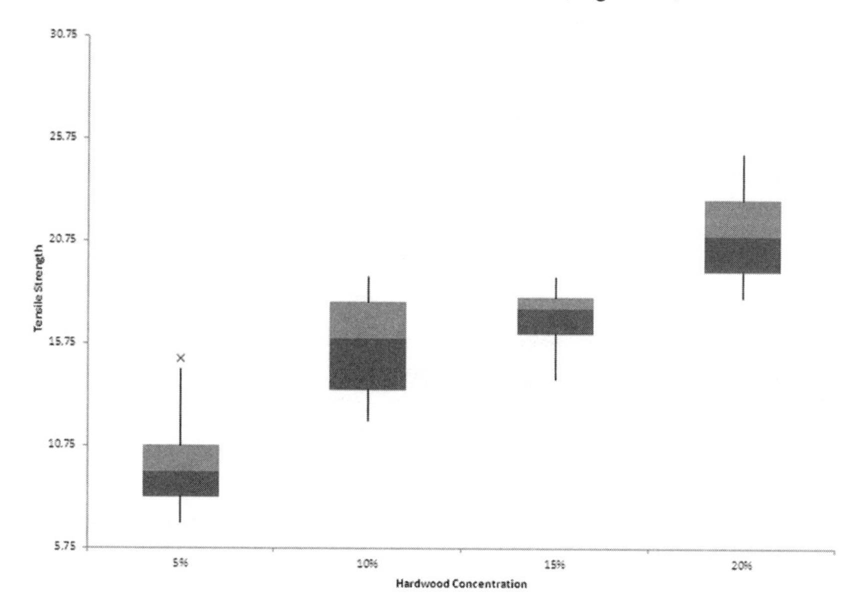

Figure 12.6 Box and whisker plot of tensile strength.

(height of the box) is smallest at15 percent hardwood and largest at 10 percent.

Remember that our customer wants a tensile strength of 15 pounds per square foot. A hardwood concentration of 15 percent clearly exceeds this level, and 10 percent is also above it. From Figure 12.5, we can see that the averages for 10 and 15 percent are over 15 pounds per square foot. How can we test to see whether they are statistically the same? We could run an ANOVA on just those two columns, or we could use a *t*-test.

T-Test for Means

We use *t*-tests to evaluate whether the means of one or two samples are the same or different. There are several types of *t*-tests:

- Two-sample *t*-test assuming *equal* variances
- Two-sample *t*-test assuming *unequal* variances
- Paired two-sample *t*-test for means
- Single-sample *t*-test

Two-Sample *t*-Test Assuming Equal Variances

Because we know from Levene's test that the variances are statistically the same (i.e., equal), we can use a *t*-test to compare the means. Just select the two columns of data and run the QI Macros two-sample *t*-test *assuming equal variances* (Fig. 12.7).

Again, the *p*-values are greater than 0.05, which means that the averages are statistically the same. Thus we can go with 10 percent hardwood to achieve our customer's requirements and save money.

	A	B	C	D	E	F
1	10%	15%	t-Test: Two-Sample Assuming Equal Variance	α	0.05	
2	12	14	Equal Sample Sizes			
3	17	18		10%	15%	
4	13	19	Mean	15.67	17	
5	18	17	Variance	7.867	3.2	
6	19	16	Observations	6	6	
7	15	18	Pooled Variance	5.533		
8			Hypothesized Mean Difference	0		
9			df	10		
10			t Stat	-0.982		
11			P(T<=t) one-tail	0.175		Accept Null Hypothesis because p > 0.05 (Means are the same)
12			T Critical one-tail	1.812		
13			P(T<=t) two-tail	0.349		Accept Null Hypothesis because p > 0.05 (Means are the same)
14			T Critical Two-tail	2.228		

Figure 12.7 *T*-test assuming equal variances.

Conduct an *F*-Test to Determine Whether Variances Are Equal

If we didn't know that the variances were equal, we could have run an *F*-test (Fig. 12.8) to compare the two samples. Again, the *p*-values are less than 0.05, so the variances are equal.

	A	B	C	D	E	F	G	H	I
1	10%	15%	F-Test Two-Sample for Variances	α	0.05				
2	12	14							
3	17	18		10%	15%				
4	13	19	Mean	15.67	17				
5	18	17	Variance	7.867	3.2				
6	19	16	Observations	6	6				
7	15	18	df	5	5				
8			F	2.46					
9			P(F<=f) one-tail	0.173	0.346	Two-tail			
10			F Critical one-tail	5.05	7.15	Two-tail			
11			One-tail	Accept Null Hypothesis because p > 0.05 (Variances are the same)					
12			Two-tail	Accept Null Hypothesis because p > 0.05 (Variances are the same)					

Figure 12.8 *F*-test two-sample for variances.

Two-Sample *t*-Test Assuming Unequal Variances

If the variances were not equal (i.e., F-test *p*-value < 0.05), we could have used the QI Macros to select the two-sample *t*-test *assuming unequal variances* (Fig. 12.9).

	A	B	C	D	E	F
1	10%	15%	t-Test: Two-Sample Assuming Unequal Variances	α	0.05	
2	12	14	Equal Sample Sizes			
3	17	18		10%	15%	
4	13	19	Mean	15.67	17	
5	18	17	Variance	7.867	3.2	
6	19	16	Observations	6	6	
7	15	18	Hypothesized Mean Difference	0		
8			df	8		
9			t Stat	-0.982		
10			P(T<=t) one-tail	0.177		Accept Null Hypothesis because p > 0.05 (Means are the same)
11			T Critical one-tail	1.860		
12			P(T<=t) two-tail	0.355		Accept Null Hypothesis because p > 0.05 (Means are the same)
13			T Critical Two-tail	2.306		

Figure 12.9 *T*-test two-sample assuming unequal variances.

QI Macros Stat Wizard for One-Click Answers

Nonstatisticians can't always remember all the rules to follow when using statistical tools. There are cumbersome decision trees to orient one's thinking, but rather than having to think your way through the forest of decision trees, wouldn't it be great if the decision tree was already coded into the software? In the QI Macros, it is! Because the QI Macros Stat

Wizard already knows what data you've selected, it can choose the most likely statistical analysis tools and run them for you.

The Stat Wizard will look at your data to determine whether it has one, two, or more columns of data. Then, depending on the number of columns and whether the data has decimals or integers, it will run the following statistics:

- One column:
  - Descriptive statistics
  - One-sample *t*-test for means
- Two columns:
  - Descriptive statistics
  - *F*-test for variances
  - *T*-test for means (assuming equal or unequal variances depending on the *F*-test)
  - Chi-square table for independence (if data is integers)
  - Fisher's test for 2 × 2 tables
  - Regression analysis
- Three+ columns:
  - Descriptive statistics (Fig. 12.10)
  - ANOVA for means
  - Levene's test for variances
  - Chi-square table for independence (if data are integers)

"Descriptive Statistics" provides the most common statistics about the data (i.e., mean, median, mode, etc.) and evaluates whether the data is normal or nonnormal. It also charts the data in four ways:

- Histogram
- Box plot
- Confidence intervals
- Probability plot (for normality)

"Descriptive Statistics" sets the stage for ANOVA, *F*-tests, *t*-tests, and other statistics. You may not need everything the Stat Wizard gives you, but you can always delete the rest. When the Stat Wizard finishes its analysis, it will give the results in plain English (Fig. 12.11).

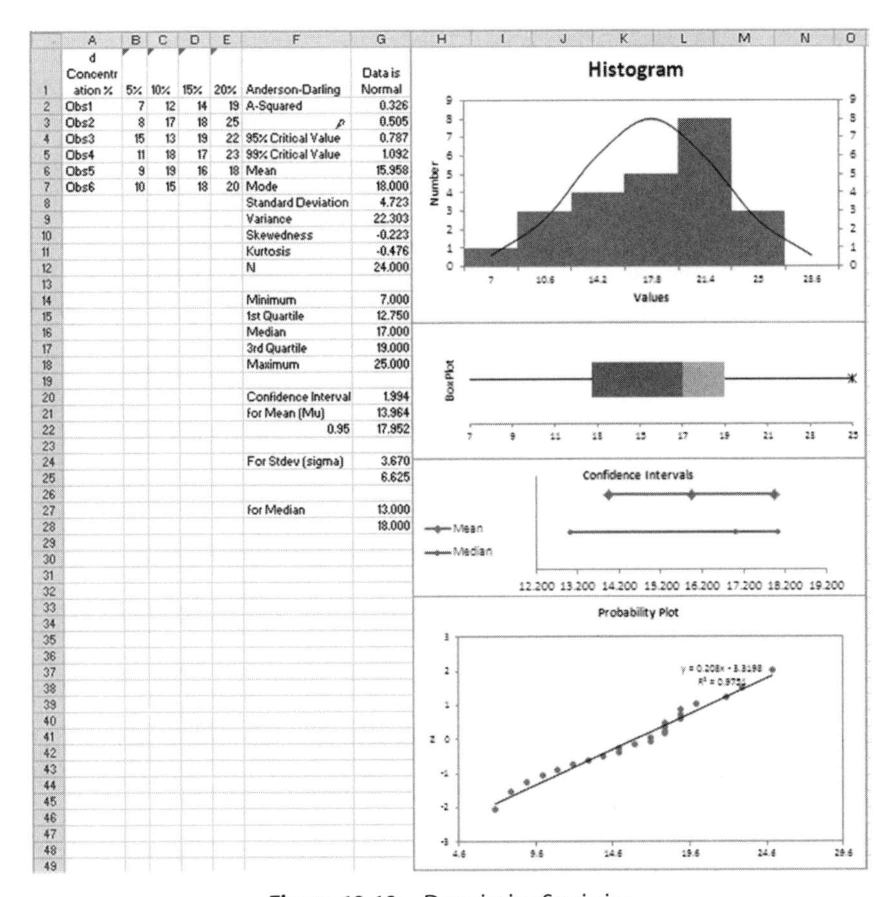

	A	B	C	D	E	F	G
1	d Concentration %	5%	10%	15%	20%	Anderson-Darling	Data is Normal
2	Obs1	7	12	14	19	A-Squared	0.326
3	Obs2	8	17	18	25	p	0.505
4	Obs3	15	13	19	22	95% Critical Value	0.787
5	Obs4	11	18	17	23	99% Critical Value	1.092
6	Obs5	9	19	16	18	Mean	15.958
7	Obs6	10	15	18	20	Mode	18.000
8						Standard Deviation	4.723
9						Variance	22.303
10						Skewedness	-0.223
11						Kurtosis	-0.476
12						N	24.000
13							
14						Minimum	7.000
15						1st Quartile	12.750
16						Median	17.000
17						3rd Quartile	19.000
18						Maximum	25.000
19							
20						Confidence Interval	1.994
21						for Mean (Mu)	13.964
22						0.95	17.952
23							
24						For Stdev (sigma)	3.670
25							6.625
26							
27						for Median	13.000
28							18.000

Figure 12.10 Descriptive Statistics.

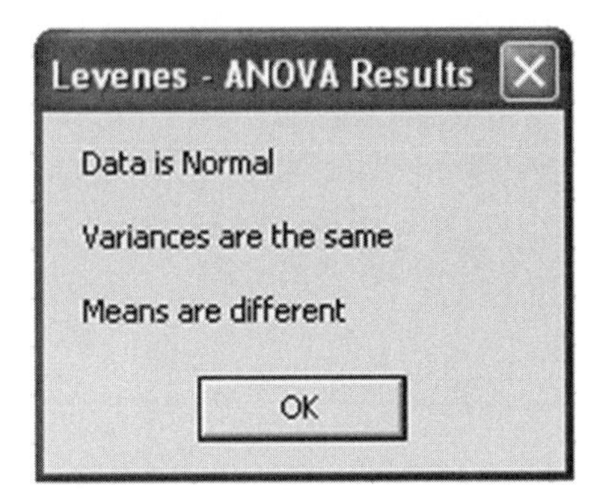

Figure 12.11 Stat Wizard results.

One-Sample *t*-Test

Sometimes we aren't comparing two samples but rather a single sample against a requirement. Let's suppose that we make light bulbs and want to know whether they meet the requirement for 2,500 hours of life. The bulb life data is shown in Figure 12.12. We can use the QI Macros one-sample *t*-test to find out if we're meeting the requirement (Fig. 12.13). Because the *p*-value is less than 0.05, the means are different. The average life of the bulbs we tested was 2,836 hours, so our bulbs *exceed* the 2,500-hour requirement. (*Note:* Excel does not have a one-sample *t*-test.)

	A
1	Bulb Life (Hours)
2	3150
3	3033
4	2862
5	2827
6	3124
7	2669
8	2364
9	2575
10	3161
11	2570
12	2860
13	2423
14	2843
15	3134
16	2959

Figure 12.12 One-sample *t*-test data for bulb life.

	A	B	C	D	E	F	G	H
1	Bulb Life (Hours)	**t-Test 1-sample**						
2	3150	Test Mean	2500					
3	3033	Confidence Level	0.95					
4	2862	N	15					
5	2827	Average	2836.93		Test Stdev	p 1-sample Stdev		
6	3124	Stdev	266.109		266.1088571	0.899		
7	2669	SE Mean	68.709					
8	2364	T	-4.904					
9	2575	TINV	1.76131					
10	3161	p - One sided	0.00012	Reject Null Hypothesis because p < 0.05 (Means are Different)				
11	2570	p - two sided	0.00023	Reject Null Hypothesis because p < 0.05 (Means are Different)				
12	2860							
13	2423							
14	2843							
15	3134							
16	2959							

Figure 12.13 Bulb life *t*-test one sample.

Weibull Analysis

Weibull analysis is another way to look at the bulb failure data. Weibull analysis is especially useful for nonnormal failure data. The QI Macros Weibull Histogram will give a chart of the failure rates and a couple of special metrics—shape and scale (Fig. 12.14). The histogram is skewed

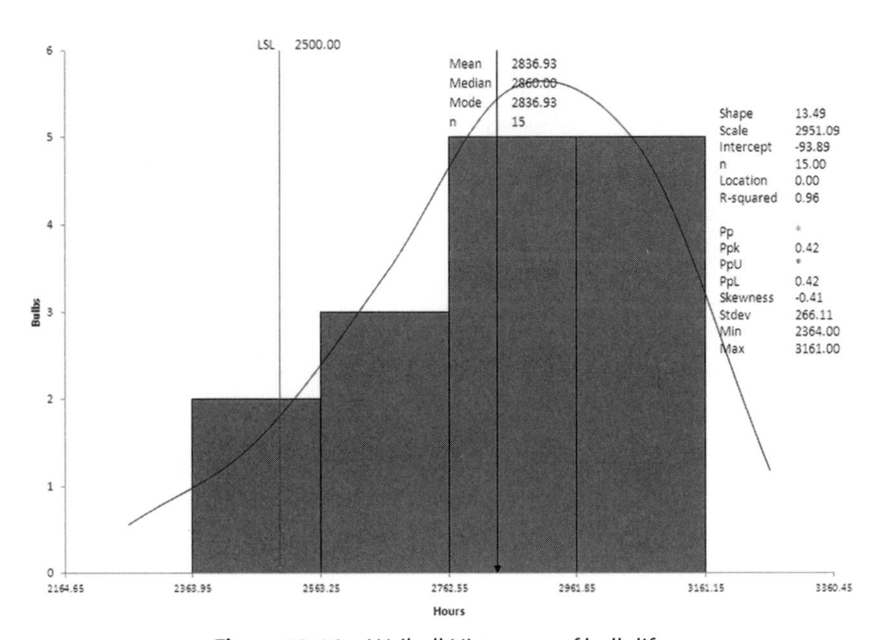

Figure 12.14 Weibull Histogram of bulb life.

toward the right (more failures later). And at least one bulb failed before the 2,500-hour requirement. (*Note:* Weibull Histograms are not included in Excel, only the QI Macros.)

The *shape* metric (13.49) is greater than 1.0, which tells us that the failure rate is *increasing* with time. A shape less than 1.0 tells us that the failure rate is decreasing. A shape of 1.0 means that the failure rate is *constant*, neither rising nor falling. The *scale* metrics tells us that two-thirds of the bulbs will have failed by 2,951 hours.

Customer Service: One-Sample *t*-Test Example

Let's say that you want to know whether wait times at a bank are not greater than five minutes. Observers collect wait times. This gives us the data we need to test the hypothesis (Fig. 12.15). Using the QI Macros one-sample *t*-test, we find that the wait times are statistically the same as the target (Fig. 12.16). The *p*-values are greater than 0.05, and the average is 4.465 minutes.

We can say that we are 95 percent confident that wait times are five minutes. We could run a Weibull Histogram on the data (Fig. 12.17) to discover that the *shape* is 1.07 (i.e., flat). The *scale* tells us that two-thirds of customers will have been served within 5.21 minutes.

F	G
Customer Wait Time at a Bank (minutes)	
4	Target =5
0	
0	
0	
10	
1	
3	
10	

Figure 12.15 Customer wait times at a bank.

	A	B	C	D	E	F	G	H	I
1	Customer Wait Time at a Bank (minutes)	t-Test 1-sample							
2	4	Test Mean	5						
3	0	Confidence Level	0.95						
4	0	N	43						
5	0	Average	4.46512		Test Stdev	p 1-sample Stdev			
6	10	Stdev	5.34678		5.34678	0.942			
7	1	SE Mean	0.81538						
8	3	T	0.656						
9	10	TINV	1.68195						
10	4	p - One sided	0.2577	Accept Null Hypothesis because p > 0.05 (Means are the same)					
11	2	p - two sided	0.5154	Accept Null Hypothesis because p > 0.05 (Means are the same)					

Figure 12.16 Customer wait times *t*-test one-sample.

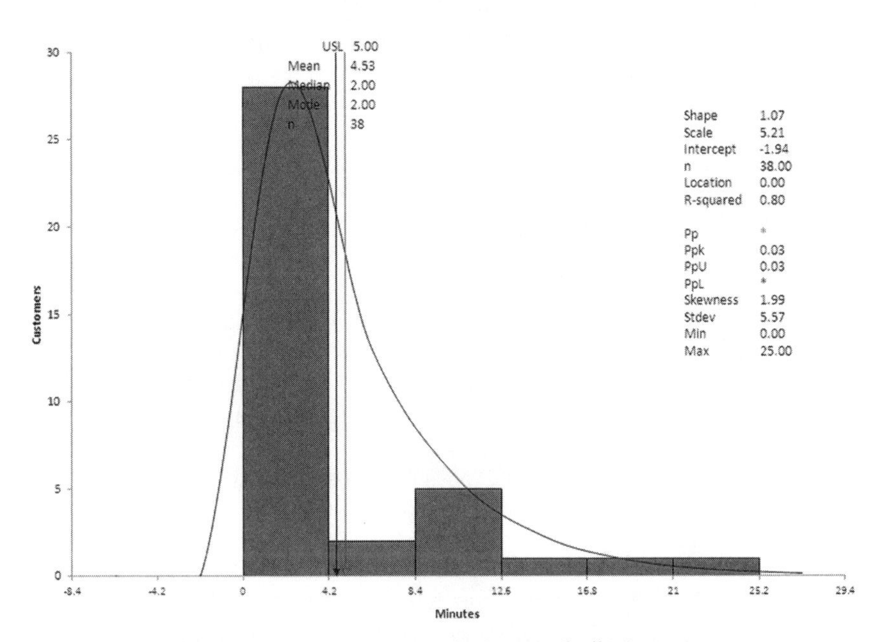

Figure 12.17 Customer wait time Weibull Histogram.

ANOVA, *t*-tests, and *F*-tests are the most common statistical tools used to fine-tune the delivery of a product or service. There are a few more, less commonly used tools.

Paired Two-Sample *t*-Test for Means

A paired *t*-test compares two *related* sets of data. It helps to determine whether the means (i.e., averages) are different from each other. An example might include test results before and after training (these are paired because the same student produces two results).

The same would be true of weight loss. If a diet claims to cause more than a 10-pound loss over a six-month period, you could design a test using several individuals' before and after weights. The samples are "paired" by each individual. You might want to know whether the diet truly delivers greater than a 10-pound weight loss.

Now conduct a test with several individuals, and enter the data into Excel. Then use the QI Macros paired two-sample t-test to select the paired two-sample *t*-test. In most *t*-tests, we want to know whether the means are the same. In this case, we're looking for a difference of 10 pounds, so we're going to enter a *hypothesized mean difference of 10* to get the result (Fig. 12.18). With the 10-pound difference, the means are the same. Thus the diet did produce a 10-pound weight loss.

	A	B	C	D	E	F	G
1	Before Diet	After Diet	Diff	t-Test: Paired Two Sample for Means	α	0.05	
2	213.4	200.1	13.3				
3	225.0	216.4	8.6		*Before Diet*	*After Diet*	
4	217.0	195.6	21.4	Mean	211.65	201.125	
5	183.7	175.0	8.7	Variance	144.0106667	175.1713	
6	197.2	201.3	-4.1	Observations	16	16	
7	223.6	214.8	8.8	Pearson Correlation	0.583731792		
8	224.2	215.7	8.5	Hypothesized Mean Difference	10		
9	215.2	200.7	14.5	df	15		
10	202.4	211.7	-9.3	t Stat	0.182		
11	217.7	216.1	1.6	P(T<=t) one-tail	0.429		Accept Null Hypothesis because p > 0.05 (Means are the same)
12	221.0	208.5	12.5	T Critical one-tail	1.753		
13	219.9	188.4	31.5	P(T<=t) two-tail	0.858		Accept Null Hypothesis because p > 0.05 (Means are the same)
14	205.4	206.4	-1.0	T Critical Two-tail	2.131		

Figure 12.18 Paired two-sample *t*-test of weight loss.

Two-Factor ANOVA

One-factor ANOVA is fine for most things, but what if you have three formulas and four different temperatures (Fig. 12.19). What is the best formula and temperature? You might run an ANOVA study with two or more *replications*.

Then, using the QI Macros, run a two-factor ANOVA with replication (Fig. 12.20).

Here the *p*-values are all less than 0.05, so the means are statistically different for temperatures and formulas. Based on what the customer wants, we can look for the best fit of formula and temperature to satisfy his or her requirements.

	A	B	C	D
1	Temperature	1-15	2-15	3-15
2	15	130	150	138
3	15	74	159	168
4	15	155	188	110
5	15	180	126	160
6	70	34	136	174
7	70	80	106	150
8	70	40	122	120
9	70	75	115	139
10	125	20	25	96
11	125	82	58	82
12	125	70	70	104
13	125	58	45	60

Figure 12.19 ANOVA two-factor data.

F	G	H	I	J	K	L	M
Anova: Two Factor With Replication	α	0.05					
SUMMARY	1-15	2-15	3-15	Total			
15							
Count	4	4	4	12			
Sum	539	623	576	1738			
Average	134.75	155.75	144	144.8333			
Variance	2056.917	656.25	674.6667	1004.515			
70							
Count	4	4	4	12			
Sum	229	479	583	1291			
Average	57.25	119.75	145.75	107.5833			
Variance	556.9167	160.25	508.25	1838.992			
125							
Count	4	4	4	12			
Sum	230	198	342	770			
Average	57.5	49.5	85.5	64.16667			
Variance	721	371	371.6667	659.0606			
Total							
Count	12	12	12	36			
Sum	998	1300	1501	3799			
Average	83.16667	108.3333	125.0833	105.5278			
Variance	2360.879	2447.515	1279.174	2218.485			
ANOVA							
Source of Variation	*SS*	*df*	*MS*	*F*	*P-Value*	*F crit*	
Sample	39118.72	2	19559.36	28.96769	0.000	3.354131	Reject Null Hypothesis because p < 0.05 (Means are Different)
Columns	10683.72	2	5341.861	7.911372	0.002	3.354131	Reject Null Hypothesis because p < 0.05 (Means are Different)
Interaction	9613.778	4	2403.444	3.559535	0.019	2.727765	Reject Null Hypothesis because p < 0.05 (Means are Different)
Within	18230.75	27	675.213				
Total	77646.97	35					

Figure 12.20 ANOVA two-factor results.

Tukey Quick Test for Means in Nonnormal Data

A Tukey quick test is like a *t*-test, but it can handle *non-parametric* (i.e., nonnormal) data. It helps to determine whether the means are the same or different from each other.

Tukey's quick test can be used when:

- There are two *unpaired* samples of similar size that overlap each other. The ratio of sizes should not exceed 4:3.
- One sample contains the highest value, and the other sample contains the lowest value. One sample cannot contain both the highest and lowest values, nor can both samples have the same high or low value.

Tukey Quick-Test Example

Using the Tukey quick test in the QI Macros Non-Parametric Tools (Fig. 12.21), it's easy to conduct the test. Because the end count = 9, we can be 98

	A	B	C	D	E
1	**Sample 1**	Sample 2	Sorted Combined Samples	Tukey Quick Test	
2	**15**	16.3	**15**	Total End Count	Confidence %
3	**16.5**	18.8	**15**	2	50%
4	**17.3**	15.8	**15**	3	63%
5	**15.3**	17.1	**15.1**	4	75%
6	**15**	17.9	**15.3**	5	84%
7	**15.1**	17.4	15.8	6	91%
8	**15**	16.7	16.3	7	95%
9	**17.6**	17.3	**16.5**	8	97%
10	**17.4**	17.5	**16.7**	9	**98%**
11	**16.7**	18.7	16.7	10	99%
12		19.5	17.1	End Count Table	
13			**17.3**	Top End Count	5.0
14			17.3	Bottom End Count	4.0
15			**17.4**	Total End Count	9.0
16			17.4	Signficant? (Y/N)	Yes
17			17.5	Confidence? (%)	98%
18			**17.6**	*p*	0.018

Figure 12.21 Tukey quick test data.

percent confident that the means are statistically *different*. (*Note:* If the data violates any of the rules, the template will not calculate the Tukey quick test.)

Chi-Square Tests in Excel

There are different types of chi-square tests.

- Chi-square goodness-of-fit tests (using Excel's "Chitest" function)
- Chi-square test of a contingency table (QI Macros Chi-Square)
- Fisher's exact test for 2 × 2 tables (use Fisher's exact test for 2 × 2 tables)

Chi-Square Goodness-of-Fit Test

A chi-square goodness-of-fit test evaluates the probabilities of multiple outcomes.

Las Vegas Dice Example

Let's say that you want to know whether a six-sided die is fair or unfair. Now test 120 rolls of the die, and enter the data into Excel (Fig. 12.22, cells A23 through C29). Then, in an empty cell, begin typing the formula "=chitest (." Excel will prompt for the observed and expected ranges. Use your mouse to select the "Observed" (B24 through B29) and "Expected" ranges (C24 through C29). Put a comma between the two and a parenthesis at the end, and hit "Return."

The chi-square test will calculate the probability (i.e., p-value) of all sides being equal. Because the p-value is 4.55759×10^{-5} (0.0000456), which is dramatically lower than our alpha value of 0.05, we can say that the die is fair.

D24 =CHITEST(B24:B29,C24:C29)

	A	B	C	D
23	Las Vegas Dice	Observed	Expected	
24	1	10	20	4.55759E-05
25	2	25	20	
26	3	30	20	
27	4	20	20	
28	5	30	20	
29	6	5	20	

Figure 12.22 Las Vegas dice.

Chi-Square Test of a Contingency Table

A chi-square test can evaluate whether two things are independent of or related to each other. We've all taken surveys and probably wondered what happened to the data. A chi-square test of a contingency table helps to identify whether there are differences between two or more demographics. Consider the following example.

Men versus Women

Imagine asking male and female patients if they agree, disagree, or are neutral about a given topic (e.g., customer satisfaction). How will we know if they have the same or differing opinions?

Conduct the survey, and enter the number of responses into Excel (Fig. 12.23, cells A1 through C4). As you can see, men seem to agree more than the women do, but is it statistically different?

	A	B	C	D	E	F
1		Men	Women	Total	**Chi-Sq**	**16.16492**
2	Agree	58	35	93	***p***	**0.000309**
3	Neutral	11	25	36	α	0.05
4	Disagree	10	23	33	Variables are Related	
5	Total	79	83	162		
6						
7		Men	Women	Contribution		
8	Agree	3.527434	3.357437			
9	Neutral	2.447961	2.329987			
10	Disagree	2.306632	2.195469			

Figure 12.23 Men and women.

Select the data, and use the QI Macros chi-square test. (*Note:* This is not part of Excel's tools.) The chi-square test macro will calculate the results. Because the *p*-value (0.000309) is less than 0.05, we can say that men and women have related views on the subject.

Fisher's Exact Test of a 2 × 2 Table

A Fisher's exact test evaluates small 2 × 2 tables better than a chi-square test because it calculates the *exact* probability. A Fisher's exact test of a 2 × 2 table helps to identify whether there are differences between two or more demographics. Consider the following example.

Men versus Women Dieting: Fisher's Exact Test Example

Imagine asking men and women whether they are dieting. How will we know if one sex diets more than the other? Conduct a survey, and enter the number of responses into Excel (Fig. 12.24, cells A1 through C3). As you can see from the data, men seem to diet less than the women do, but is it statistically significant?

	A	B	C	D	E	F
1		Men	Women	Total	**Fishers**	
2	Dieting	1	9	10	***p 2-Tail***	**0.00275946**
3	Not Dieting	11	3	14	Chi-Sq	10.9710254
4	Total	12	12	24	*p*	0.00092527

Figure 12.24 Fisher's exact test.

Use the QI Macros Fisher's exact test. (*Note:* This is not part of Excel.) Fisher's exact test will calculate the exact test statistic and the chi-square statistic. Because the *p*-value, 0.00276, is less than 0.05 and 0.01, we can say that women diet more than men.

Regression Analysis

If you think one thing causes another, you can use regression analysis to confirm or deny that they are related. Use the scatter diagram or regression analysis tool under the QI Macros Statistical Tools menu to validate your suspicions.

To run regression analysis using the QI Macros:

1. Select the labels and data.
2. In Excel, select the QI Macros regression analysis.
3. Input the confidence level (e.g., 0.95).
4. Evaluate the R^2 (>0.80 is a good fit).
5. Evaluate the *F*- and *p*-values.
6. Get the equation for the fitted data.
7. Use the equation to predict other values.

Door-to-Balloon Time versus Mortality Rate Example

Imagine, for example, that a hospital wanted to know whether a heart patient's door-to-balloon (DTB) time of less than 90 minutes affects acute myocardial infarction (AMI) mortality (Fig. 12.25). Use the QI Macros to run a regression analysis at confidence level of 0.95 (Fig. 12.26).

	A	B
1	Patients Door to Balloon < 90 minutes	Acute Mycardial Infarction mortality
2	57.1%	1.7%
3	51.5%	1.9%
4	89.5%	1.4%
5	74.2%	1.3%
6	84.6%	1.4%
7	96.0%	0.7%
8	91.7%	1.1%
9	91.3%	0.8%
10	92.6%	0.7%
11	92.4%	0.8%

Figure 12.25 DTB versus AMI data.

	D	E	F	G	H	I	J	K	L
1	SUMMARY OUTPUT								
2									
3	*Regression Statistics*								
4	Multiple R	0.8746							
5	R Square	0.765	Goodness of Fit < 0.80						
6	Adjusted R Square	0.7356							
7	Standard Error	0.0022							
8	Observations	10							
9									
10	ANOVA								
11		*df*	*SS*	*MS*	*F*	*P-value*			
12	Regression	1	0.000127858	1E-04	26.042	0.001			
13	Residual	8	3.92776E-05	5E-06					
14	Total	9	0.000167136					Confidence Level	
15							0.95		0.99
16		*Coeffici*	*Standard Erro*	*t Stat*	*P-value*	*Lower 95%*	*Upper 95%*	*Lower 99*'	*Upper 99%*
17	Intercept	0.0313	0.003880559	8.055	0.000	0.0223089	0.0402061	0.018237	0.044278
18	Patients Door to Balloon < 90 minute	-0.024	0.004649501	-5.103	0.001	-0.034449	-0.013005	-0.03933	-0.008126
19									
20	y = 0.031 -0.024*Patients Door to Balloon < 90 minutes								

Figure 12.26 DTB versus AMI regression.

Analysis

Because R^2 (0.765) is less than 0.80, as it is in this case, there is not a good fit to the data. This means that 76.5 percent of the variation in AMI can be explained by the percent of patients with DTB times of less than 90 minutes.

Because the p-value of 0.001 is less than 0.05, we can say that DTB and AMI are related. This relationship also could be investigated using a scatter diagram (Fig. 12.27). Based on this chart, mortality *declines* as the percent of patients completing angioplasty rises.

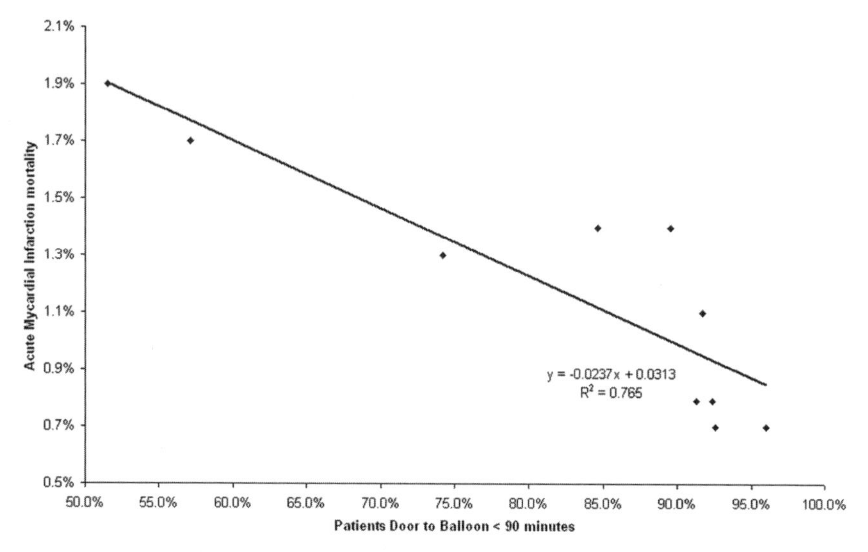

Figure 12.27 DTB versus AMI scatter.

Multiple Regression Analysis

The purpose of multiple regression analysis is to evaluate the effects of two or more independent variables on a single dependent variable. Select 2 to 16 columns with the dependent variable (i.e., the result) in the first (or last) column. Imagine, for example, that we want to know whether customer perception of quality varies with various aspects of geography and shampoo characteristics: foam, scent, color, or residue. Use the QI Macros to do multiple regression analysis on this data at the 95 percent level (Fig. 12.28, matrix-plot.xls).

H	I	J	K	L	M	N	O	P
SUMMARY OUTPUT		Force Constant to Zero						
		FALSE						
Regression Statistics								
Multiple R	0.894							
R Square	0.800	Goodness of Fit >= 0.80						
Adjusted R Square	0.745							
Standard Error	2.206							
Observations	24							
ANOVA								
	df	*SS*	*MS*	*F*	*P-value*			
Regression	5	350.4359369	70.08718739	14.40738733	0.000			
Residual	18	87.56406307	4.864670171					
Total	23	438					Confidence Level	
						0.95		0.99
	Coefficients	*Standard Error*	*t Stat*	*P-value*	*Lower 95%*	*Upper 95%*	*Lower 99%*	*Upper 99%*
Intercept	90.1921828	4.046988933	22.286244	0.000	81.68977456	98.69459105	78.54317	101.8412
Region	-3.859117116	1.042809786	-3.700691313	0.002	-6.04997918	-1.668255053	-6.86078	-0.85745
Foam	1.816893637	0.35820658	5.072195036	0.000	1.064329538	2.569457736	0.785817	2.84797
Scent	1.034684641	0.922490779	1.121620579	0.277	-0.903396568	2.97276585	-1.62065	3.690019
Color	0.232672236	0.708123097	0.328575974	0.746	-1.255039185	1.720383658	-1.80562	2.270962
Residue	-4.00148838	0.811858697	-4.928799056	0.000	-5.707140211	-2.29583655	-6.33838	-1.6646

Figure 12.28 Multiple regression.

Analysis

Because R^2 is greater than or equal to 0.80, as it is in this case, there is a good fit to the data. Looking at the *p*-values for each independent variable "Region," "Foam," and "Residue," we see that they are less than alpha (0.05), so we can say that these variables have an impact on quality. "Scent" and "Color" *p*-values are greater than 0.05, so we can say that there is no correlation.

Conclusion

Although these tools are extremely useful for deeper analysis of your data, most breakthrough improvement practitioners aren't ready to dive into them until they have a firm grasp of the basic measurement and improvement processes. If you want to learn more about these, consider *Advanced Statistics DeMYSTiFieD* (Stephens 2004).

Exercises

Use the test data in c:\qimacros\testdata\anova.xls to practice using these tools.

BIBLIOGRAPHY

Arthur, Jay, *Lean Six Sigma Demystified*. New York: McGraw-Hill, 2011.

Arthur, Jay, *Lean Six Sigma for Hospitals*. New York: McGraw-Hill, 2011.

Berry, Leonard Eugene, *Management Accounting DeMYSTiFieD*. New York: McGraw-Hill, 2006.

Bossidy, Larry, and Ram Charan, *Execution*. New York: Crown Business, 2002.

Christiansen, Clayton, *The Innovator's Dilemma*. Boston: Harvard Business School Press, 2000.

Cyr, Jay, et al., "Sustaining and Spreading Reduced Door-to-Balloon Times for ST-Segment Elevation Myocardial Infarction Patients," *Joint Commission Journal on Quality and Patient Safety*, June 2009, pp. 297–306.

Downes, Larry, and Chunka Mui, *Unleashing the Killer App*. Boston: Harvard Business School Press, 1998.

Farzad, Roben, "The Toyota Enigma," *BusinessWeek*, July 10, 2006, p. 30.

Gladwell, Malcolm, *The Tipping Point*. Boston: Little Brown, 2002.

Godin, Seth, *Unleashing the Ideavirus*. New York: Hyperion, 2001.

Goldratt, Eliyahu M., and Jeff Cox, *The Goal: A Process of Ongoing Improvement*. Great Barrington, MA: North River Press, 1984.

Hall, Kenji, "No One Does Lean Like the Japanese," *BusinessWeek*, July 10, 2006, pp. 40–41.

Heath, Dan, and Chip Heath, *Decisive*. New York: Crown Business, 2013.

Kaplan, Robert S., and David P. Kaplan, *The Balanced Scorecard*. Boston: Harvard Business School Press, 1996.

Kaplan, Robert S., and David P. Kaplan, *The Strategy-Focused Organization.* Boston: Harvard Business School Press, 2001.

Kim, Christopher S., et al., "Implementation of Lean Thinking: One Health System's Journey," *Joint Commission Journal on Quality and Patient Safety*, August 2009, pp. 405–413.

Liker, Jeffrey, *The Toyota Way.* New York: McGraw-Hill, 2004.

McCorvey, J. J., "The Race Has Just Begun," *Fast Company*, September 2013, pp. 66–76.

Moore, Geoffrey, *Crossing the Chasm.* New York: Harper Business, 1999.

Morse, Gardiner. An Interview with Atul Gawande, "Health Care Needs a New Kind of Hero," *Harvard Business Review*, April 2010, Vol. 88, Iss. 4; p.60(2).

Nicole, Adrian, "A Gold Medal Solution," *Quality Progress*, March 2008, pp. 45–50.

Pelczarski, Kathryn, and Cynthia Wallace, "Hospitals Collaborate to Prevent Falls," *Patient Stafey & Quality Healthcare*, November–December 2008, pp. 30–36.

Powers, Donna, and Mary Paul, "Healthcare Department Reduces Cycle Time and Errors," *Six Sigma Forum Magazine*, February 2008, pp. 30–34.

Raynor, Michael E., and Mumtaz Ahmed, "Three Rules for Making a Company Truly Great," *Harvard Business Review*, Vol. 91, Iss. 4; p. 108–117, April 2013; available at: http://hbr.org/2013/04/three-rules-for-making-a-company-truly-great/ar/1.

Roam, Dan, *The Back of the Napkin.* New York: Penguin, 2009.

Roam, Dan, *Blah, Blah, Blah.* New York: Penguin, 2011.

Rogers, Everett, *Diffusion of Innovations*, 4th ed. New York: Free Press, 1995.

Stauk, George, and Thomas Hout, *Competing Against Time.* New York: Free Press, 1990.

Stock, Greg, "Taking Performance to a Higher Level," *Six Sigma Forum Magazine*, May 2002, pp. 23–26.

Tufte, Edward, *Envisioning Information*. Cheshire, CT: Graphic Press, 1990.

Tufte, Edward, *Visual Explanations*. Cheshire, CT: Graphic Press, 1997.

Tukey, J. W., "A Quick, Compact, Two-Sample Test to Duckworth's Specifications," *Technometrics*, Vol. 1, No. 1, pp. 31–48, February 1959.

Wennecke, Gette. "Kaizen: Lean in a Week," August 2008; available at: www.mlo-online.com.

Widner, Tracy, and Mitch Gallant, "A Launch to Quality," Quality Progress, Febrary 2008, pp. 38–43.

Womack, James P., and Daniel T. Jones., *Lean Thinking*. New York: Simon & Schuster, 1996.

LEARNING RESOURCES

Breakthrough Improvement Video Training:
www.breakthrough-improvement-excel.com

QI Macros Video Training:
www.qimacros.com/training/videos/

QI Macros "How To" Articles:
www.qimacros.com/free-resources/excel-tips/

Excel Tips and Techniques:
www.qimacros.com/training/excel-videos/

QI Macros 90-Trial:
www.qimacros.com/breakthrough

Free Lean Six Sigma Yellow Belt Training:
www.lssyb.com

INDEX